T0092849

# Physical Oceanography of Continental Shelves

# Physical Oceanography
# of Continental Shelves

K. H. Brink

Princeton University Press
Princeton and Oxford

Published by Princeton University Press

41 William Street, Princeton, New Jersey 08540
99 Banbury Road, Oxford OX2 6JX
press.princeton.edu

Library of Congress Cataloging-in-Publication Data

Names: Brink, Kenneth H., author.
Title: Physical oceanography of continental shelves / K.H. Brink.
Description: Princeton : Princeton University Press, 2023. |
    Includes bibliographical references and index.
Identifiers: LCCN 2022045750 (print) | LCCN 2022045751 (ebook) |
    ISBN 9780691236452 (hardback) | ISBN 9780691236469 (ebook)
Subjects: LCSH: Continental shelf. | Continental slopes. |
    Continental margins. | Physical oceanography.
Classification: LCC GC85 .B75 2023 (print) | LCC GC85 (ebook) |
    DDC 551.46/8—dc23/eng20230323
LC record available at https://lccn.loc.gov/2022045750
LC ebook record available at https://lccn.loc.gov/2022045751

British Library Cataloging-in-Publication Data is available

Editorial: Ingrid Gnerlich and Whitney Rauenhorst
Production Editorial: Kathleen Cioffi
Text and Cover Design: Wanda España
Production: Jacqueline Poirier
Publicity: William Pagdatoon
Copyeditor: Barbara Liguori

Jacket image: Sediment off the Yucatan Peninsula. NASA image courtesy the MODIS Rapid Response Team at NASA GSFC

This book has been composed in Minion Pro, 10/13

Printed on acid-free paper. ∞

Printed in the United States of America

10  9  8  7  6  5  4  3  2  1

To my many generous teachers, past and present

# Contents

# Preface

This volume is an attempt to present core material about continental shelf physical ocean-ography in a way that begins with the basics and yet can be used as a starting point for exploring frontier areas in depth. It grew out of teaching a one-semester second-year graduate-level course in the Woods Hole/MIT Joint Program. Given this anticipated level, the reader should have some knowledge of quantitative physical oceanography, although I have attempted to provide enough background material to allow a broader audience to find its way. Even though this work is conceived as a textbook, only a few exercises (the appendix) are provided for the reader. This is largely because informative exercises are difficult to create and require extensive vetting in practice. I have on hand only this small collection. I have found, however, that a really valuable teaching tool is to assign each student an important paper to study, to present verbally, and to be able to answer critical questions.

The structure of this volume is motivated by the notion that although the global coastal ocean is tremendously diverse, locations differ mainly in how the various under-lying processes interact. With that in mind, emphasis is on quantitatively exploring some of the important physical mechanisms. There are many options about which topics to include. While I believe the right choices were made, I readily admit that the content reflects my own experiences and priorities. Other scientists would perhaps have made other choices. In any case, numerous topics receive little or no attention, even though they are undeniably interesting and important. These omissions include ice-related pro-cesses (e.g., Straneo and Cenedese, 2015), estuaries (e.g., MacCready and Geyer, 2009; Geyer and MacCready, 2014; Bruner de Miranda et al., 2017), turbulent mixing (e.g., Gregg, 2021), hydraulics (Pratt and Whitehead, 2007), practical ocean numerical modeling (e.g., Pinardi et al., 2017), and the surf zone (e.g., Komar, 1998). These refer-ences are but a sampling of the many excellent books or reviews that help make my omissions less blameworthy.

Another set of choices had to be made about how to include ocean observations. No book of this sort could fail to include information about the real ocean and how it moti-vates and tests models. The approach here is to discuss measurements in the context of appropriate models (or vice versa), rather than to have separate chapters, sorted perhaps regionally, on observations. The difficulty with using the present approach is that some cohesiveness is lost in terms of characterizing ocean observations in a particular region. Fortunately, substantial, albeit now somewhat dated, resources exist (Robinson and Brink, 1998, 2006) that synthesize regional coastal ocean observations with nearly global coverage.

On a personal basis, I entered coastal physical oceanography in the 1970s, a time of exciting growth. I have been fortunate enough to have known and sometimes worked with a range of outstanding and pioneering figures, all of whom have been wonderful teachers in one way or another. I regret that I missed a couple of the earliest practitioners, notably June Pattullo and Henry Bryant Bigelow, but I am very grateful for the many I have known.

Finally, there are many people to thank both for my own education and for their direct or indirect contributions to this volume. Of those who have helped me, I single out my mentors: George Veronis, John Allen, Bob Smith, and Bob Beardsley. I could not imagine my life without them. For this volume I especially appreciate Steve Lentz, who was particularly helpful during the formative stages of this process. Jamie Pringle, Sasha Yankovsky, and Jim Price provided critical readings of drafts and motivated numerous improvements. Input along the way from Allan Clarke, Steve Elgar, André Paloczy, Chris Piecuch, and three anonymous reviewers is also gratefully acknowledged. Of course, any errors or shortcomings are attributable only to me. Finally, this volume would not exist if it were not for the many students I have interacted with, who motivated this effort and who taught me all sorts of things along the way.

## Table of Consistently Used Symbols

| Symbol | Definition | Defining equation |
|---|---|---|
| $A$ | Eddy viscosity | (2.3.4a) |
| $Bu$ | Burger number | (4.4.7) |
| $c$ | Wave speed (subscripts provide more specifics) | |
| $c_D$ | Drag coefficient for quadratic stress | (3.6.7) |
| $d$ | Bottom frictional parameter | (3.7.4) |
| $D$ | Free-surface divergence parameter | (4.2.10) |
| $Ek$ | Ekman number | (2.3.6) |
| $f$ | Coriolis parameter $= 2\Omega \sin \phi$, where $\phi$ is latitude | (2.3.1) |
| $g$ | Acceleration due to gravity | |
| $h$ | Spatially variable depth | |
| $h_{ML}$ | Surface mixed-layer thickness | |
| $H$ | Constant water depth | |
| $i$ | Square root of $-1$ | |
| $(k, l)$ | Horizontal wavenumbers in the $(x, y)$ directions | |
| $K$ | Eddy diffusivity | (2.3.4b) |
| $N$ | Buoyancy frequency | (2.3.7) |
| $p$ | Pressure | |

| Symbol | Definition | Defining equation |
|---|---|---|
| $r$ | Radial coordinate | |
| $R_B$ | Bulk Richardson number | (3.5.7) |
| $Re$ | Reynolds number | (2.3.2) |
| $Ri$ | Gradient Richardson number | (2.3.7) |
| $Ro$ | Rossby number | (2.4.7) |
| $s$ | Slope Burger number | (3.7.6) |
| $S$ | Salinity | |
| $t$ | Time | |
| $T$ | (with no subscripts) Temperature | |
| $(u, v, w)$ | Velocity components in the offshore, alongshore, and vertical directions | |
| $(x, y, z)$ | The offshore, alongshore, and vertical coordinates | |
| $\alpha$ | Bottom slope | |
| $\beta$ | North-south gradient of $f$ | (2.4.8) |
| $\delta_D$ | Dirac delta function | (4.4.3) |
| $\delta_E$ | Scale thickness of a constant eddy viscosity Ekman layer | (3.2.4) |
| $\delta_{nm}$ | Kronecker delta | (5.4.11) |
| $\Xi$ | Cross-sectional $(x, z)$ area | |
| $\pi$ | 3.14159 . . . | |
| $\rho$ | Fluid density | |
| $\sigma_F$ | Bottom resistance coefficient | (3.6.10) |
| $\Sigma$ | Summation | |
| $\omega$ | Wave frequency | |
| $\Omega$ | Earth's rotation rate | |

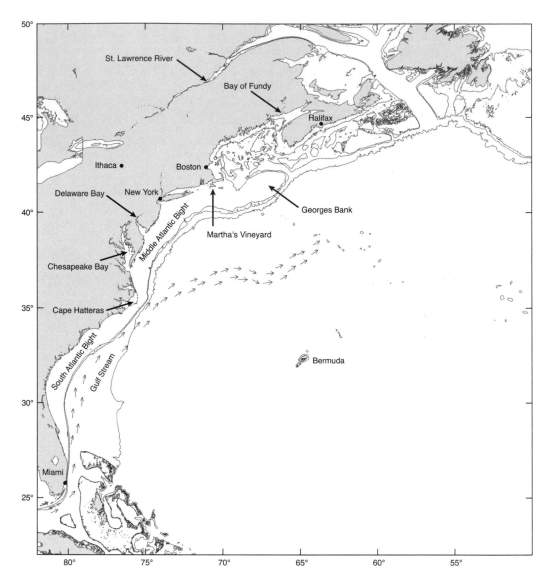

**MAP 1.** Western North Atlantic. The 100, 200, and 2000 m isobaths are shown. Created using GEBCO 2020 gridded digital data.

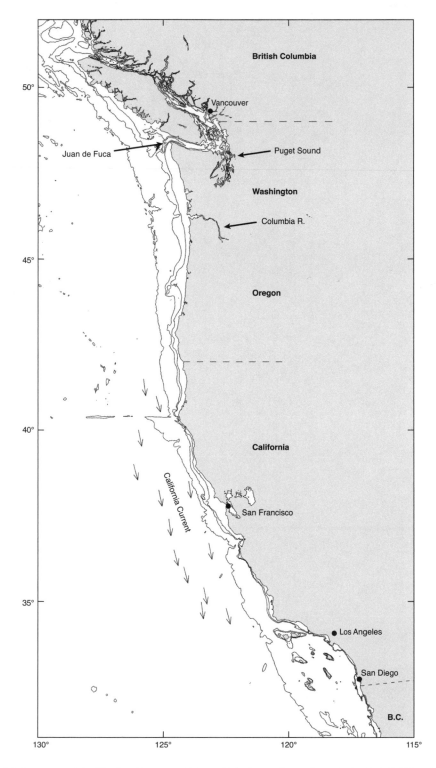

**MAP 2.** Eastern North Pacific. The 100, 200, and 2000 m isobaths are shown. Created using GEBCO 2020 gridded digital data.

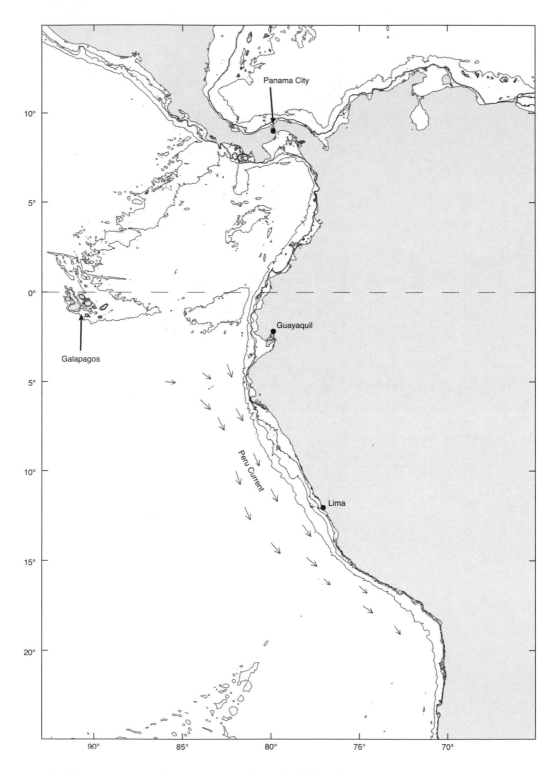

**MAP 3.** Eastern Tropical Pacific. The 100, 200, and 2000 m isobaths are shown. Created using GEBCO 2020 gridded digital data.

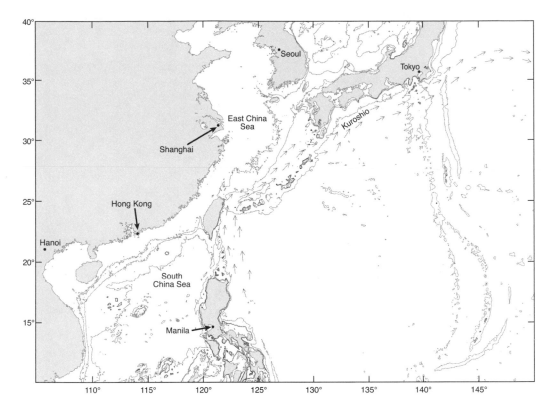

**MAP 4.** Western North Pacific. The 100, 200, and 2000 m isobaths are shown. Created using GEBCO 2020 gridded digital data.

# Physical Oceanography
# of Continental Shelves

# 1
# Introduction

## 1.1    What Does "Coastal" Mean?

There are many good definitions of the coastal ocean. An inclusive one might be: all the salty water adjoining continents and islands where the water is shallower than some arbitrary depth, like 1000 m. This definition would include estuaries, the surf zone, continental shelves, and continental slopes, but it needs to be supplemented by the inclusion of large enclosed freshwater bodies, such as the Laurentian Great Lakes, as coastal as well. Chapter 2 provides more detail on terminology, but, for now, the two important threads are proximity to land and the occurrence of large fractional depth changes.

## 1.2.    Why Is the Coastal Ocean Important?

The coastal ocean, because it adjoins land, is the most visible and heavily used part of the world's ocean. Most people may never see the deep open ocean aside from perhaps through an airplane window. Given its proximity to land, the coastal environment is used heavily for recreation (swimming, boating, diving, fishing) and enjoyed aesthetically. People are particularly aware of the coastal environment.

This nearness to land also imposes a good deal of pressure on the coastal ocean. For a range of reasons, human populations concentrate near the coast (Figure 1.1). In the United States, 39% of the population lives in counties that adjoin the ocean, according to the U.S. Census Bureau, and the number is substantially higher when land adjoining the Great Lakes is included. Globally, according to NASA, 40% of the population lives within 100 km of the ocean. Thus, settlement patterns concentrate both the observers and stressors of the coastal environment.

The connection to land extends well beyond coastal settlements, however. Rivers drain much of the world's continental surface into the ocean. The outflow waters carry sediments (Figure 1.2), an assortment of dissolved chemicals, and other materials. Humanity, of course, can affect any of these quantities, for example, through agricultural practices, damming, and waste disposal.

**FIGURE 1.1.** View of the United States at night: a composite of nighttime images from 2012. The lighting, which reflects population levels, distinctly outlines the coastline around much of the United States. From NASA Earth Observatory/ NOAA NGDC.

Some people call the coastal environment "the dirty little bathtub ring around the world's ocean." There is, in fact, a good deal of truth to this quip. The coastal ocean is very productive biologically for several reasons, some of which will be explained in the following chapters. Figure 1.3 shows a long-term average estimate of chlorophyll concentration in the upper ocean. Chlorophyll is a commonly used proxy for the concentration of the microscopic plants (*phytoplankton*) that are the base of ocean food chains. The important point in this figure is that the highest concentrations (light shades) all occur in the coastal ocean. The pale bands along the equator are an expression of processes analogous to (but more diffuse than) those acting near a coast. Viewed from a ship, the biologically productive waters often look green, brown, or blackish (that dirty bathtub ring), while beautiful Mediterranean-blue waters are actually an expression of the relative absence of life and are more representative of midocean conditions (the darker shades in Figure 1.3) than coastal ones. When phytoplankton are plentiful, the food web is stimulated even up to the level of fish, so it is no surprise that many of the world's most productive fisheries (such as for anchovies off Peru) are in the coastal ocean. Palomares and Pauly (2019), for example, estimated that the coastal ocean accounts for 55% of the global fish catch even though it represents only 3% of the ocean surface.

**FIGURE 1.2.** Daytime visual satellite image of the Connecticut River outflow during a flood stage after a hurricane (September 11, 2011). The river's suspended sediments make the muddy outflow plume quite visible as lighter shades. The outer part of Long Island is in the lower part of the image. NASA Earth Observatory image by Robert Simmon, using *Landsat 5* data from the US Geological Survey.

The abundance of coastal fisheries is not simply a matter of plentiful phytoplankton. Rather, fish need to find their prey (smaller animals), and these need to find their prey, and so on, through the food web. The locations and movements of all the levels of predator and prey are, of course, mitigated by currents and mixing. In the extreme, fish eggs generally drift entirely at the mercy of the physical setting. The success of a species depends

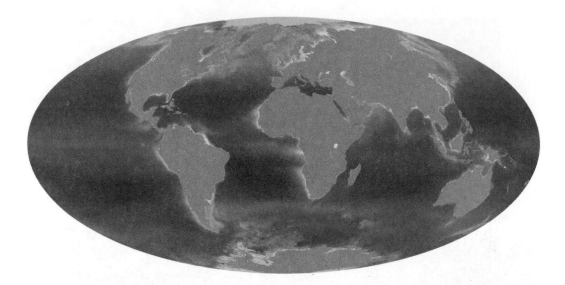

**FIGURE 1.3.** Global average ocean chlorophyll concentration (1997–2007) estimated from satellite images of ocean color. Lighter shades represent higher concentrations. Chlorophyll is often used as a proxy for phytoplankton biomass. Courtesy of NASA Earth Observatory.

upon how well suited the animal's evolved behavior is to its actual environment. Because the physical system is variable over time and space, it follows that ecosystem structure varies as well, even putting aside strictly biological mechanisms. This array of physical-biological couplings has motivated a good deal of research over the years, one illustrative example being described by Wiebe et al. (2002).

Biological activity in the coastal realm is not always benevolent, however. The rivers that empty into the ocean sometimes carry a heavy content of dissolved nutrients that frequently originate from excessive use of agricultural fertilizers, among other sources. These nutrients stimulate plankton growth, which eventually leads to sinking of organic material over the continental shelf. The sinking material decays, consuming oxygen. The upshot is the seasonal appearance of areas where near-bottom shelf waters are so nearly depleted in dissolved oxygen (*hypoxic*) that fish can no longer survive there. Shelf hypoxia is a growing problem globally (e.g., Rabalais et al., 2009). Another unfortunate biological effect occurs when there are blooms of plankton species that are harmful to humans and coastal animal species. These harmful algal blooms (HABs) (e.g., Anderson et al., 2012), like hypoxia, are strongly affected by aspects of the physical setting such as water column density stratification. HABs are often called "red tides," even though they only occasionally redden the water, and tidal currents are almost irrelevant.

Fishing, of course, is only one of many human activities in the coastal ocean. Shipping has been important since classical times and is increasingly so today. Transportation issues are especially pressing in areas near ports where safe and efficient navigation calls for coordination and a knowledge of the environment. Even small gains in efficiency

**FIGURE 1.4.** Oil platforms off the coast of California. Courtesy of the U.S. Bureau of Ocean Energy Management.

can represent important savings, given the cost of operating a large vessel. Inevitably, heavy usage leads to occasional calls for search and rescue operations, which clearly benefit from a knowledge of winds and currents. The coastal environment remains important for naval operations, since these often involve shipping (protection or prevention), or various offensive and defensive deployments near land.

There are other valuable resources in the coastal ocean. Oil and gas exploitation (Figure 1.4), which will be important for the foreseeable future, calls for knowledge about currents to enable safe and efficient operations. Petroleum, of course, is not the only coastal energy source: wind energy, as well as other renewables, stand to become increasingly important over the coming years. The continental shelf setting is particularly attractive because "land use" is perhaps not as difficult an issue as ashore and because winds are often stronger over the water than over the nearby land. Further, there are other minerals to be extracted besides fuels: sand and even diamond extraction both lead to disruption of the coastal ocean seafloor.

This cursory listing of practical concerns is far from complete and entirely without detail. The object is to emphasize how the coastal ocean is disproportionately (by area) important to our society, for a diversity of reasons. The following chapters will rarely touch explicitly on practical concerns, but applications are always nearby and will be a strong motivator to many readers.

## 1.3.    What Makes the Coastal Ocean Different?

As simple as it may sound, probably the most distinctive aspect of the coastal ocean is the presence of a boundary. The blockage inhibits cross-shelf velocity $u$ while having little direct effect on the alongshore current component $v$. This inhibition contributes heavily to the tendency for $u$ to be much weaker than $v$ over the continental shelf, especially on time scales longer than a few days (e.g., Figure 1.5; Lentz and Fewings, 2012). The resulting anisotropy, in turn, leads to momentum balances strikingly dependent on orientation, hence different from the more nearly isotropic open ocean. Also, the coastal barrier disrupts the cross-shelf upper-ocean transport associated with Earth's rotation and alongshore winds, effectively creating an outsized near-surface flow divergence near shore. (Such a divergence, albeit much more broadly distributed, is a key link in wind driving of the open ocean; e.g., see Gill, 1982, his chapter 12). This large divergence, in turn, means that wind-driven currents are more energetic over the shelf than in the deeper ocean. Consequently, coastal currents vary (on time scales longer than tidal periods) over periods typical of the weather: about 2–15 days. In contrast, subtidal mid-ocean currents typically fluctuate with periods defined by mesoscale eddies: many tens of days.

Another defining aspect of the coastal ocean is the large degree of bathymetric variation. Across the continental margin, bottom-depth changes are comparable to the depth itself. Given the reluctance of slowly varying flows on a rotating planet to cross *isobaths* (contours of constant depth), depth changes reinforce the boundary's tendency to make flow, on time scales longer than tidal, follow isobaths (Figure 1.5). In addition, the depth changes, through a funneling effect, tend to amplify tidal currents as the water gets shallower. Beyond that, there is the possibility of tidal resonances in the presence of a coastal wall. Strong tides, in turn, lead to a range of secondary effects, including enhanced nutrient delivery and thus biological activity. Finally, the very shallowness of the shelf's water column means that a given forcing (such as a wind stress or heat flux) is relatively more effective than in the deep ocean, where the effect might be distributed over a far greater vertical extent.

In most places in the ocean, there are turbulent boundary layers near the surface and bottom. These are typically 10–50 m thick. In the deep (thousands of meters) ocean, these boundary layers do not occupy a very large fraction of the water column. But on the continental shelf, where waters are typically 150 m deep or shallower, the boundary layers occupy a substantial part, and sometimes all, of the water column across the shelf. Consequently, turbulent mixing and dissipative processes play a particularly important role in coastal phenomena.

Finally, the coastal environment is where the continent meets the ocean. What flows out of rivers passes into coastal waters, creating distinctive alongshore buoyant currents which carry chemicals and materials that originated inland. Thus, the coastal setting is the ocean's contact zone with the terrestrial environment, and it is often here that outflows are diluted to concentrations typical of the broader open ocean.

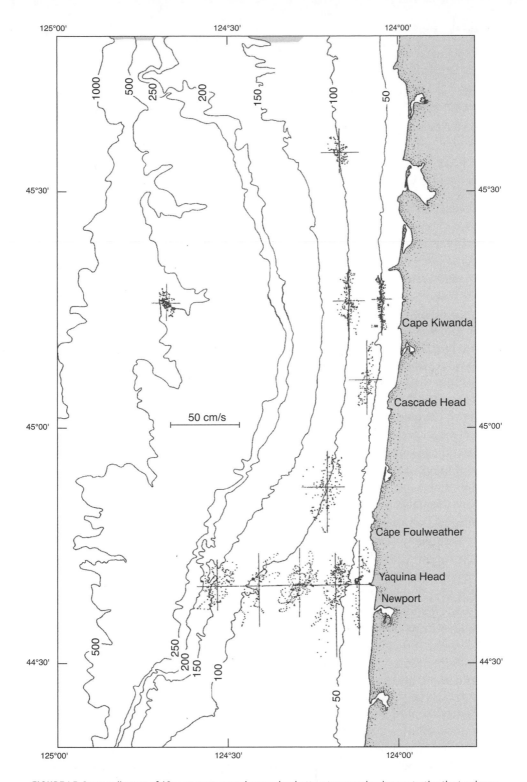

**FIGURE 1.5.** Scatter diagram of 40 m currents superimposed on bottom topography, demonstrating the tendency for fluctuating currents to follow isobaths. At each location, the current meter records were smoothed to remove tides and higher-frequency motions, and then the mean currents were removed. Next, at regular time intervals, the velocity vector was plotted as a dot relative to its mooring location (crosshairs). The measurements were made during the summers of 1971 and 1972. Note how the clouds of points are stretched out and aligned roughly parallel to isobaths, especially near the coast. Depths are in meters. Adapted from Kundu and Allen (1976).

This summary is deliberately terse. However, all these concepts are dealt with in detail in the following chapters. After reading those, the reader can return to this chapter with an understanding of the physics underlying these sweeping statements.

## 1.4.    The Common Theme: Cross-Shelf Exchange

The focus of coastal ocean research repeatedly turns to cross-shelf exchange. There are many reasons why this is so. Coastal upwelling (section 4.1), which is so important for biological productivity, is the transport of deeper waters onshore, up into the sunlight, and then offshore at the surface. In many other cases, cross-shelf gradients—of salinity, for example—reflect inputs from land spreading offshore. It is, in fact, quite common to find that cross-shelf gradients greatly exceed alongshore gradients, especially on regional and larger scales. In the face of these anisotropic property distributions, a relatively weak cross-shelf flow can still effect substantial net transports. Many interesting problems, then, such as how the Gulf Stream affects shelf waters, boil down to cross-shelf exchange. Thus, it has been a dominant focus of continental shelf research for a half century or more.

If cross-shelf exchange has been a priority for so long, why is it still a going concern? There are at least two answers. One is that we all have an understandable tendency to deal first with the most visible variability in a data set. Because tides are obvious to see and important for practical reasons, they naturally received attention from very early times, making them arguably the first topic treated in physical oceanography. Aristotle and Laplace were both engaged with this well before the twentieth century. Coastal upwelling was understood to be important by the early twentieth century and thus motivated perhaps the first major coordinated field program of continental shelf research (Coastal Upwelling Ecosystems Analysis: CUEA; see Hartline, 1980). But this program made some of its most important physical contributions on wind-driven motions and coastal-trapped waves: processes that dominate alongshore currents on time scales of days or longer. Although CUEA certainly dealt with the cross-shelf flows that structure upwelling, the more salient findings came from attacking the strongest then-unexplained variability: that involving fluctuating alongshore currents. Great progress was made, but it was not always where originally intended. A second reason for slow progress on cross-shelf transport is that it is simply a difficult problem. There is no single answer that applies everywhere. Cross-shelf currents are generally weaker than alongshore (on time scales longer than tidal), so they can be hard to separate from alongshore currents (as silly as that may appear, this is a real issue over a bottom with curvy isobaths). Further, the cross-shelf flow often has much shorter natural length scales than does the alongshore flow, so it is far more difficult to map out a coherent field.

Altogether, there are good reasons that cross-shelf exchanges remain a central focus of continental shelf research. But new understandings, new ideas, and new tools will continue to enable new progress.

# 2
# Some Basic Concepts

## 2.1.  Introduction

It is useful to introduce some vocabulary, assumptions, and concepts that recur throughout the following chapters. This material may already be familiar to many readers, but the goal is to have a common starting point.

## 2.2.  The Setting

The continental margins occur around all major landmasses, so the individual features described here are commonly found in various guises. The diagrams in Figure 2.1 introduce some key features, but, admittedly, simplistically. Most strikingly, the coastline and the isobaths are all drawn perfectly straight. While it is sometimes true that actual coastlines are relatively straight, and that alongshore depth variations are usually less dramatic than those in the offshore direction, curvature and contortion are usually the rule in reality (e.g., Figure 1.5 or the maps on pages xiv–xvii).

The broad features of continental margins are often determined by plate tectonics. Specifically, tectonically active margins (Figure 2.1, lower panel), such as off Peru or Japan, are characterized by oceanic plates sinking below continental crust. In these regions, there is often a terrestrial coastal mountain range, the overall continental margin is relatively narrow, and there is frequently an adjoining oceanic *trench* in deep water. Active margins tend to have a good deal of seismic activity as well. Passive margins occur far from spreading centers or subduction zones. These are often comparatively wide regions (Figure 2.1, upper panel) and often adjoin land with relatively gentle terrain. The continental margins of Argentina or the eastern United States are good examples.

The *continental shelf* (Figure 2.1) is a relatively gently sloping area adjoining the shoreline and extending offshore to meet the much more steeply sloping *continental slope*. The boundary between these regions is often a well-defined *shelf break* where the bottom slope abruptly increases. While shelves can be as narrow as a few kilometers (as near San Diego, California), they are typically 20–100 km wide and sometimes substantially more,

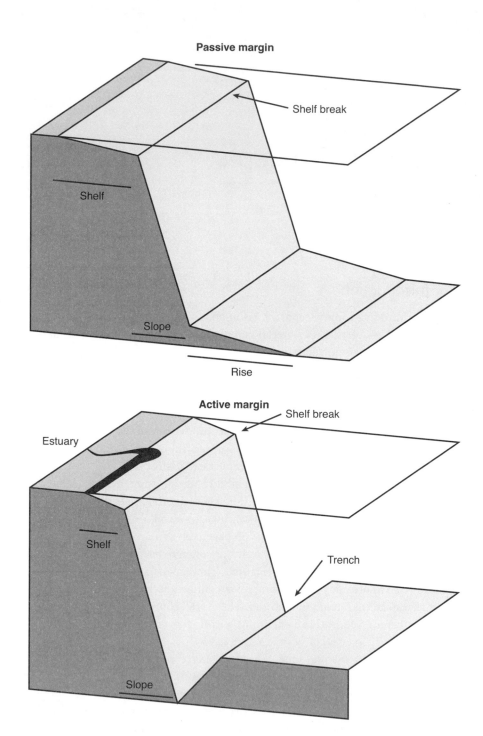

**FIGURE 2.1.** Schematic diagrams of coastal ocean geometries, defining commonly used terms. Upper: a tectonically passive margin; Lower: a tectonically active margin. The darkest shading on the active margin represents estuarine outflow and its consequent buoyancy current.

as in the Bering Sea or north of Siberia. Very often, the outer edge of the shelf is in about 100–200 m of water, but some shelf edges are as shallow as about 30 m or as deep as 450 m. A typical shelf bottom slope might be about 0.002. Shelves along tectonically passive margins tend to be relatively wide, while those at active margins are often narrower. In some cases, shelf topography is fairly smooth, but there are very rugged shelves, punctuated by basins and rises, such as off Nova Scotia. Further, shelves are often crossed by submarine valleys that might extend across the entire shelf (such as the Hudson Shelf Valley offshore of New York City or the Juan de Fuca Channel off the west coast of North America) or by deep, dramatic canyons at the shelf edge, as occur frequently south of New England, for example.

Continental slopes are the real topographic boundary between the deep sea and the coastal ocean. Typically, slope depths might range from about 100 m down to thousands of meters and do so over offshore scales of only a few tens of kilometers. Representative bottom slopes are 0.01 to even 0.1 in extreme cases. Sometimes, the continental slope directly adjoins a relatively flat abyssal plane (depth typically 4000–5000 m), but in other cases, there is a gently sloping *continental rise* (Figure 2.1, upper panel) that lies between the lower slope and the more nearly flat abyss. A good example is the passive margin south of New England where the rise occupies a depth range of about 2500–4000 m and is a few hundred kilometers wide.

The coastal ocean's landward boundary has its own set of features. Shorelines, where the land meets the ocean, can be sandy, rocky, or even vegetated. Adjacent to coasts that are exposed to waves coming in from offshore there is generally a very dynamic *surf zone*, which is defined by the presence of breaking waves. Often, this corresponds to waters shallower than about 3 m, but the actual defining isobath can vary dramatically over time scales as short as hours in response to the properties of incoming waves. Just offshore of the surf zone lies the *inner shelf*, which is defined by waters shallow enough that turbulent (boundary layer) effects are important throughout the water column. Typically, the inner shelf might extend out to the 30 m isobath, but this bounding depth can vary by an order of magnitude, depending on currents and surface fluxes. Finally, breaks in the coastline allow *estuaries* (Figure 2.1, lower panel), which are the contact zone between freshwater rivers and the salty ocean. Usually, estuaries are surrounded by land and are the site of diverse tidal and mixing processes that determine the properties of the brackish water that eventually exits onto the shelf. The relatively fresh water that leaves an estuary often turns and flows alongshore as a *buoyancy current*, which is less salty than the ambient shelf waters.

## 2.3.    Turbulence

The Navier-Stokes equations describe the momentum balances that govern motions in the ocean (or in any other conventional fluid, for that matter). They describe physics ranging from flow in a human capillary or motions around a flying golf ball up to scales representative of global ocean circulation. This range expresses the wonderful applicability of the equations, but it also represents a problem. Consider flow in the ocean: at the same time there are regional-scale currents (scale of hundreds of kilometers), wave motions (scales of meters to many kilometers), and disorderly small-scale motions (centimeters up to many kilometers). In practical terms, one cannot hope to resolve all these motions in detail simultaneously; no computer is that powerful, nor is any observing system that capable. Thus, simplifications need to be made by deciding at the outset what scales are of greatest interest and then either seeking conditions for neglecting other scales or finding ways to express the effects of unresolved scales on the scale of interest. For example, if one is concerned with mean midshelf circulation (scale of tens of kilometers), one might be able to neglect the effects of passing waves (scales of meters to kilometers), but one might need to account for the dissipative processes at the smallest scales (centimeters to meters). The following discussion deals with disorderly small-scale processes and how they interact with larger scales of motion.

As a starting point, consider an incompressible fluid in a spatial domain that is small relative to a rotating planet's radius, so the curvature of Earth's surface can be neglected. The advantages of this system, compared with using the spherical coordinates appropriate to Earth as a whole, is that it is relatively easy to visualize, and it is often used in coastal oceanography where the scales of interest are limited. The approximations involved in deriving this simplified system are discussed well elsewhere (e.g., Vallis, 2017). The momentum balances express how the acceleration of a fluid particle (first four terms of each equation) is balanced by the Coriolis acceleration (the terms involving $f$), pressure differentials, and viscous stress divergences:

$$u_t + (uu)_x + (vu)_y + (wu)_z - fv = -\frac{1}{\rho}p_x + v(u_{xx} + u_{yy} + u_{zz}), \tag{2.3.1a}$$

$$v_t + (uv)_x + (vv)_y + (wv)_z + fu = -\frac{1}{\rho}p_y + v(v_{xx} + v_{yy} + v_{zz}), \tag{2.3.1b}$$

$$w_t + (uw)_x + (vw)_y + (ww)_z = -\frac{1}{\rho}p_z + v(w_{xx} + w_{yy} + w_{zz}) - g. \tag{2.3.1c}$$

The continuity equation is

$$u_x + v_y + w_z = 0. \tag{2.3.1d}$$

The variables $(u, v, w)$ are the velocity components in the $(x, y, z)$ directions (with $z$ upward), $\rho$ is the density of water, $p$ is pressure, $f$ is the Coriolis parameter ($= 2\Omega \sin \varphi$, where $\Omega$ is Earth's rotation rate, and $\varphi$ is latitude), $g$ is the acceleration due to gravity, and $v$ is the kinematic molecular viscosity coefficient. Subscripted independent variables $(x, y, z, t,$ where $t$ is time) indicate partial differentiation. This simplification to Cartesian coordinates means that, realistically, north-south scales more than about 1000 km should not be considered. Further, only the component of Earth's rotation in the local vertical direction is included here (e.g., Gill, 1982, his chapter 7.4), an approximation that is generally excellent, especially away from the equator. Yet, this set still describes motions on scales ranging from millimeters up to hundreds of kilometers. More simplification is desirable.

The multiple nonlinear terms on the left-hand side of equations (2.3.1) allow the possibility of generating *turbulence*, a flow state that is highly disorderly, nonlinear, and ultimately dissipative. Turbulence can be contrasted with waves, which often are more orderly, have linear dynamics, and obey tightly defined relations between time and length scales. Turbulent flow is chaotic, and the nonlinearity allows the transport of properties such as heat or momentum. Comparing the magnitude of the nonlinear terms to the viscous terms in (2.3.1) yields the *Reynolds number*

$$Re = \frac{\hat{u}L}{\nu}, \tag{2.3.2}$$

where $\hat{u}$ and $L$ are representative scales for velocity and length, respectively. For example, at the smallest scales typically associated with shelf circulation, $\hat{u} = 0.01$ m/s and $L = 1$ km, along with the given $\nu = 0.9 \times 10^{-6}$ m$^2$/s (at room temperature), $Re = O(10^7)$ (read this as $Re$ is of the order of $10^7$). This value means that, at this scale, nonlinear processes overwhelmingly dominate the effects of molecular viscosity. But what does this imply?

Under controlled laboratory conditions, such as for flow in a pipe, Reynolds numbers larger than about few thousand are associated with turbulent flow; that is, the nonlinear terms dominate over dissipation, and chaotic flow occurs on smaller scales. Thus, even with weak flow on oceanic scales, $Re = O(10^7)$, oceanic flow would be inescapably associated with turbulence. The turbulence, being a function of the flow field, is unevenly distributed and intermittent in both space and time. Some locations, such as near boundaries, can experience almost continual turbulence, but other locations may experience it only in occasional bursts.

What is ocean turbulence actually like? The turbulence responsible for vertical mixing is relatively isotropic and typically cascades energy from larger-generation, $O(10$ m$)$, scales down to the smallest, $O(0.01$ m$)$ or less, scales where molecular viscosity finally dissipates the kinetic energy, and $Re = O(1)$ (i.e., molecular viscosity plays a decisive role). The resulting mixing, too, is relatively isotropic, so this scale of turbulence can contribute to lateral mixing in contexts involving shorter horizontal scales. At the largest vertical scales, stable density stratification often inhibits vertical turbulent motions, so for horizontal turbulent scales of $O(100$ m$)$ or more, the turbulent flow is constrained to move nearly horizontally. This larger-scale, relatively two-dimensional turbulence plays an important role in horizontal mixing and can occur on scales up to many tens of kilometers. But two-dimensional turbulence has properties distinctly different from isotropic smaller-scale turbulence: most notably the tendency for energy to cascade from smaller to larger horizontal scales (e.g., Vallis, 2017), so Earth's rotation is usually an important factor. It is thus fairly common for scientists who study vertical (small-scale, relatively isotropic) and horizontal (larger-scale, anisotropic) mixing to follow somewhat different approaches.

Given the range of scales involved, it is often hopeless to try to resolve turbulent motions in models or theories of many processes in the ocean. Rather, normal practice is to seek a "spectral gap," that is, to find (hope for?) a clear separation, both in space and time, between the scales of interest (say, on the horizontal scale of the shelf and with a time scale of days) and of unresolved turbulent motions, which might have a range of scales down to a few centimeters, and to seconds for the small-scale turbulence that is often responsible for vertical mixing. If there is such a scale separation, then it is reasonable to try to *parameterize* the effects of the turbulent component of the flow field. In this context, to parameterize something means to combine information on the resolved scale with physical insight about smaller scales to estimate the effects of unresolved scales on the larger scales. To be specific, one can break up variables into a slowly varying (relative to the turbulent fluctuations) "mean" part $<u>$ and a turbulent part $u'$, where $<>$ here denotes an average over the turbulent fluctuations, that is, over all space and time scales shorter than those defining the spectral gap. Thus,

$$u = <u> + u' \tag{2.3.3a}$$

or for salinity (a scalar)

$$S = <S> + S'. \tag{2.3.3b}$$

Although the average turbulent variation is zero, $<u'> = 0$, averages of products do not have to vanish (as in the simple case of the average of $\cos(t)$ compared to $\cos^2(t)$ over a cycle), so the turbulent flux of salinity, for example, is often nonzero:

$$<u'S'> \neq 0. \tag{2.3.3c}$$

It is these *eddy*[1] fluxes, and their divergences, that express the effect of turbulence on the slowly varying "mean" variables, and it is these flux divergences that need to be quantified. However, parameterizing turbulence is no simple matter and remains a very active research topic (e.g., Gregg, 2021).

There are at least two important properties of a turbulent flow field in three spatial dimensions. One is that it mixes properties such as heat and momentum; that is, normally we expect, averaging over smaller scales, turbulence to smooth out larger-scale gradients. Second, the strength of the turbulence, hence mixing, depends on the larger-scale flow field that is somehow exciting the turbulence. Although some careful arguments have been made about how to parameterize the effect of turbulence, it is perhaps simplest to assume, by analogy to molecular viscosity, that the net effect can be treated in terms of an *eddy viscosity A*. For example, the turbulent component of the momentum flux $wu$ in (2.3.1a) is represented by

$$\langle w'u' \rangle = -A <u>_z \tag{2.3.4a}$$

---

1 The term "eddy" in this context is simply taken to mean anything that varies relative to the $<>$ average. In another context, ocean eddies can be understood to be large (e.g., 100 km), roughly circular, swirling flow features in midocean. In the present context of turbulence, however, the term does not imply specific, discrete flow patterns of this sort.

That is to say that the net effect of unresolved motions can somehow be expressed in terms of properties of the resolved scales, both through $\langle u \rangle_z$ and through the eddy viscosity, which itself usually depends on properties of the larger-scale setting, such as density stratification. Similarly, for a scalar like salinity $S$

$$\langle w'S' \rangle = -K < S >_z, \tag{2.3.4b}$$

where the *eddy diffusivity* is $K$. In practice, once an eddy viscosity or diffusivity is introduced, the averaging is understood, and thus the averaging operator $<>$ is generally not shown for the larger-scale variables such as $< u > \to u$. In other words, the variables $u$, $S$, and so on, are now taken to represent only what is happening on the resolved (non-turbulent) side of the spectral gap. The form (2.3.4) is sensible in that it honors the first primary descriptor of turbulence; that is, it mixes properties in a plausible way. For example, dissolved materials are transported on average from areas with high concentrations into regions with low concentrations. Since $A$ and $K$ are generally found to be far larger than their molecular equivalents, only turbulent mixing and dissipation are generally included when considering processes on scales larger than a few meters. However, there is still the critical question of what $A$ and $K$ are in practice.

As mentioned earlier, the turbulent processes that effect vertical versus horizontal mixing differ considerably (e.g., Gregg et al., 2018, and Moum, 2021, vs. Okubo, 1971; Davis, 1985; and LaCasce, 2008). Thus, estimated vertical versus horizontal viscosities and mixing coefficients also differ considerably, with the horizontal eddy coefficients typically being orders of magnitude larger. All considered, the momentum equations (2.3.1a–c), allowing for turbulent effects (2.3.4) and using (2.3.1d), become

$$u_t + uu_x + vu_y + wu_z - fv = -\frac{1}{\rho}p_x + A_H(u_{xx} + u_{yy}) + (A_V u_z)_z, \tag{2.3.5a}$$

$$v_t + uv_x + vv_y + wv_z + fu = -\frac{1}{\rho}p_y + A_H(v_{xx} + v_{yy}) + (A_V v_z)_z, \tag{2.3.5b}$$

$$w_t + uw_x + vw_y + ww_z = -\frac{1}{\rho}p_z + A_H(w_{xx} + w_{yy}) + (A_V w_z)_z - g, \tag{2.3.5c}$$

where this form assumes, for simplicity, that $A_H$ is horizontally isotropic and spatially uniform, both assumptions which can be relaxed. Similarly, equations describing transport of a scalar take forms like

$$S_t + uS_x + vS_y + wS_z = K_H(S_{xx} + S_{yy}) + (K_V S_z)_z, \tag{2.3.5d}$$

where $K_H$ and $K_V$ are eddy diffusivities. The eddy coefficients $A_H$, $A_V$, $K_H$, and $K_V$ are functions of the flow and density fields, so the problem is far from closed at this point. The scale transfer of energy in two-dimensional turbulence often behaves differently than in isotropic three-dimensional turbulence, so even the sign of the horizontal eddy coefficients can be uncertain in some situations. Sometimes, especially in analytical theories, horizontal mixing is ignored ($A_H = K_H = 0$). In numerical models, a particularly well-resolved grid might allow the development of eddies that represent a realistic level of horizontal mixing, thus minimizing the need to parameterize this process, although some weak horizontal mixing may still be required for numerical stability in many cases.

There is less ambiguity about vertical mixing: it is undeniably important for a range of problems, most especially in turbulent layers near the surface and bottom (chapter 3). If only a qualitative result is desired, it is common practice to take $A_V$ and $K_V$ to be constant. When more realistic results are desired, there is a range of carefully constructed *turbulence closure* models (e.g., Wijesekera et al, 2003; Basdurak et al., 2021) that attempt to make quantitatively useful space- and time-dependent estimates of $A_V$ and of $K_V$. These models are often complicated and are generally shaped by a deep knowledge of turbulent flow. Their use in numerical models often leads to very credible results. And, of course, there is a middle ground of eddy viscosity forms (see chapter 3) that respect some aspects of realistic turbulence but retain some simplicity, for example, by being constant in time while remaining spatially variable.

Very often, the importance of friction is estimated by comparing the magnitude of the vertical frictional term in (2.3.5a or b) with the Coriolis term. The result is the *Ekman number*

$$Ek = \frac{A_V}{fH^2},$$
(2.3.6)

where $H$ is a depth of the entire water column, and $A_V$ is taken to be a representative value of the vertical eddy viscosity. As will become clear in chapter 3, this number can be thought of as the squared magnitude of the ratio of a turbulent boundary thickness to the overall water depth. Thus, a small Ekman number suggests that the boundary layers occupy only a small portion of the water column.

One nondimensional number is particularly important for turbulence in a stratified ocean: the gradient *Richardson number*. This is often given by

$$Ri = \frac{-g\rho_z}{\rho\hat{u}_z^2} \equiv \frac{N^2}{\hat{u}_z^2},$$
(2.3.7)

where $N^2$ is the *buoyancy frequency*[2] squared. The Richardson number compares the stabilizing tendency for density stratification to inhibit vertical motions versus the tendency for a strongly sheared flow to break down into turbulence. Thus, $Ri$ appears as a key parameter in many turbulence closure schemes. One particularly famous application of this number is in a stability condition for nonrotating stratified shear flow (Howard, 1961), where the value $Ri = 1/4$ separates stable (larger $Ri$) configurations from unstable ones (i.e., those developing Kelvin-Helmholtz instability). The Richardson number, however, appears in a range of problems on scales up to at least the ocean mesoscale, where it is a determinant of baroclinic instability (e.g., Stone, 1966). In any case, the more supercritical $1/Ri$ becomes, the stronger the shear relative to density stratification, and the stronger the ultimate growth of turbulence.

---

2 Sometimes also called the Brunt-Väisälä frequency.

## 2.4.    Some Frequently Used Approximations

A number of approximations are made so frequently in the literature that they are often not mentioned, or they are simply mentioned with no further comment. Several of these are reviewed here. Gill (1982) and Vallis (2017), for example, provide more thorough explanations and justifications.

Because density variations in liquid ocean water are always small compared with the total density, the *Boussinesq approximation* is used almost universally in oceanography. The justification for this assumption begins with the statement, quite reasonable for seawater, that the density can be written as

$$\rho = \rho_0 + \rho_V(x, y, z, t), \tag{2.4.1a}$$

where $\rho_0$ is a constant such that

$$\rho_0 \gg |\rho_V|. \tag{2.4.1b}$$

For example, even in the coastal ocean, where there can be extreme density contrasts, the range of $|\rho_V|/\rho_0$ is no more than 4% and is at least an order of magnitude less for nearly all situations outside of estuaries. Implicit in (2.4.1b) is the easily met constraint that the water depth $H$ is small enough that $|H\rho_{Vz}| \ll \rho_0$. Applying these assumptions leads simply to replacing $\rho$ with the constant $\rho_0$ in the horizontal momentum equations (2.3.5a, b), while the vertical equation (2.3.5c) can be written as

$$\rho_0(w_t + uw_x + vw_y + ww_z) = -p_z + \rho_0 A_H(w_{xx} + w_{yy}) + \rho_0(A_V w_z)_z - g(\rho_0 + \rho_V). \tag{2.4.2}$$

That is to say that density is replaced everywhere in the momentum equations by the constant background value except in the buoyancy (last) term of the vertical equation.

In addition, it is common to account for the static part of pressure by saying that $p = p_0 + p_V$, where $p_0(z)$ is associated with the constant background density

$$p_{0z} = -g\rho_0. \tag{2.4.3a}$$

So, integrating in $z$,

$$p_0 = p_{ATM} + g\rho_0\zeta - g\rho_0 z, \tag{2.4.3b}$$

where $p_{ATM}$ is the atmospheric pressure at the ocean's surface, and $\zeta$ is the free surface displacement (positive upward). Thus, the horizontal pressure gradient is

$$p_x = (p_{ATM} + g\rho_0\zeta + p_V)_x, \tag{2.4.3c}$$

$$p_y = (p_{ATM} + g\rho_0\zeta + p_V)_y. \tag{2.4.3d}$$

So that, effectively, in the horizontal momentum equations

$$p = p_{ATM} + g\rho_0\zeta + p_V, \tag{2.4.3e}$$

and (2.4.2) becomes

$$\rho_0(w_t + uw_x + vw_y + ww_z) = -p_{Vz} + \rho_0 A_H(w_{xx} + w_{yy}) + \rho_0(A_V w_z)_z - g\rho_V. \tag{2.4.4}$$

In practice, the subscript $V$ is dropped, and the main change is that the vertical equation (2.4.4) includes only the variable part of density in the last term. This simplification (removal of $g\rho_0 z$) in $p$ is made so frequently in theories that it is rarely pointed out. While atmospheric pressure needs to be accounted for in computing ocean pressure from tide gauge data, in models it is often either ignored or, with caution, absorbed into $\zeta$.

Another common assertion, used almost invariably in the following chapters, is the *hydrostatic approximation*, which requires that the vertical scales of motion are small relative to the horizontal scales. Specifically, scaling the continuity equation

$$u_x + v_y + w_z = 0 \tag{2.4.5}$$

leads to the conclusion that vertical motions are small, $O(H/L)$, relative to horizontal. Using the horizontal momentum equations (2.3.5a, b) leads to an estimate of the magnitude of pressure variations. The upshot is that the vertical equation of motion (2.3.5c) becomes, to $O[(H/L)^2]$,

$$0 = -p_z - g\rho, \tag{2.4.6a}$$

and (2.4.4) becomes

$$0 = -p_{Vz} - g\rho_V. \tag{2.4.6b}$$

The smallness of $w$ is consistent with the neglect of the Coriolis force associated with the ground-parallel rotation component, already mentioned in section 2.3. Some important classes of motion *in*appropriate for the hydrostatic approximation include three-dimensional turbulence, short internal waves, and short, high-frequency surface gravity waves (section 6.3).

In the following chapters, it is often assumed that the governing dynamics are linear, or nearly so. On a rotating planet, the natural way to justify this is to compare the nonlinear terms in the momentum equations (2.3.5a, b) with the Coriolis acceleration term, that is, to consider the ratio

$$Ro = \frac{\hat{u}}{fL}, \tag{2.4.7}$$

which is known as the *Rossby number*. One physical interpretation of this number is that it is the ratio of relative vorticity, $\eta = v_x - u_y = O(\hat{u}/L)$, to planetary vorticity ($f$). There are other criteria for judging nonlinearity as well, depending on the context. For example, in wave applications, linearity requires that the particle motions must be slow relative to the wave propagation speed; that is, $\hat{u} \ll c$, where $c$ is a typical wave propagation speed.

Earth's rotation can be ignored when the Rossby number is really large. For example, for small-scale ocean turbulence or (more familiarly) for flow in in a kitchen sink, the scale $L$ is so small that the Rossby number might be $O(10,000)$. Rotation can also be safely ignored when considering even linear motions with a time scale short compared with $f^{-1}$. For example, surface gravity waves, which are so noticeable from a ship, might have a period of about 10 s, hence frequency about 1000 times $f$, and so their propagation is not affected by rotation.

A few comments are needed about the effect of the locally vertical component of Earth's rotation, which is measured by the Coriolis parameter $f$. One effect of rotation is to further suppress the vertical velocity component beyond the geometric effect imposed by continuity (2.4.5). This suppression is enhanced by stable density stratification that imposes an energetic cost for moving parcels in the vertical. Weakened vertical velocity then has two consequences. First, the tendency toward hydrostatic balance is reinforced. Second, a vanishing vertical velocity component leads to a tendency for steady rotating flow to follow isobaths (section 10.1).

In most coastal applications, the Coriolis parameter $f$ is taken to be simply a constant or perhaps slowly varying over distance. In some situations, models allow for the latitudinal variation in $f$ by means of the "*beta-plane*" approximation, where

$$f \cong f_0 + \beta y, \tag{2.4.8}$$

$y$ is the northward coordinate, the constant $\beta = 2\Omega \cos(\varphi)/a$, $a$ is Earth's radius, and Cartesian coordinates are retained. This approximation is no longer accurate when $y$ becomes comparable to $a$. In open-ocean contexts, $\beta$, as the gradient of planetary vorticity, is often a very important parameter. As such, it sometimes arises in problems involving linkages between the coastal and open ocean. Over the continental margins, this planetary $\beta$ is often overwhelmed by topographic effects. For example, in an inviscid, uniform-density (homogeneous) ocean, the potential vorticity $\Pi = (f + \eta)/h$ (where $h$ is the water's depth) is conserved in the absence of forcing (e.g., Pedlosky, 1979, his section 3.4, or section 8.4 of this volume). Thus, as a water parcel moves about, its relative vorticity $\eta$ is strongly constrained by

$$\nabla \frac{f}{h} = \frac{\beta}{h} \boldsymbol{j} - f(\nabla h)/h^2, \tag{2.4.9}$$

(where $\boldsymbol{j}$ is a northward unit vector). Thus, $-\dfrac{f(\nabla h)}{h}$ is often called the "topographic $\beta$," because it plays the same role as the planetary vorticity gradient but is due to depth variations. Frequently, especially over the continental slope, the depth gradient is so large as to overwhelm planetary $\beta$, at least locally. Thus, the "*f-plane*," that is, $\beta = 0$, assumption is often made in coastal contexts. However, in some cases, such as coastal-trapped waves (section 5.7), $f$ is treated as a slowly varying parameter. Further, there are coastal contexts where a more nuanced accounting for Earth's curvature than the $\beta$-plane is needed (e.g., Miles, 1972), but these are exceptional.

Finally, making a few additional assumptions can justify the frequently used *shallow-water equations*. The starting point is to assume that density is constant everywhere, so that $p_z = 0$, and it is then natural to have $u_z = v_z = 0$ outside any boundary layers. Using these simplifications, depth-averaging (2.3.5a, b) and (2.4.5) leads to

$$\bar{u}_t + \bar{u}\bar{u}_x + \bar{v}\bar{u}_y - f\bar{v} = -g\zeta_x + A_H(\bar{u}_{xx} + \bar{u}_{yy}) - \frac{1}{\rho_0 h}\tau_B^x + \frac{1}{\rho_0 h}\tau_0^x \tag{2.4.10a}$$

$$\bar{v}_t + \bar{u}\bar{v}_x + \bar{v}\bar{v}_y + f\bar{u} = -g\zeta_y + A_H(\bar{v}_{xx} + \bar{v}_{yy}) - \frac{1}{\rho_0 h}\tau_B^y + \frac{1}{\rho_0 h}\tau_0^y \qquad (2.4.10b)$$

$$(h\bar{u})_x + (h\bar{v})_y + \zeta_t = 0 \qquad (2.4.10c)$$

where $(\bar{u}, \bar{v})$ is the depth averaged horizontal velocity, $\zeta$ is the free-surface height perturbation, $(\tau_0^x, \tau_0^y)$ is the surface wind stress, and $(\tau_B^x, \tau_B^y)$ is the bottom stress. Atmospheric pressure here is assumed to be constant, although this need not be assumed in
general. This shallow-water system, as written, is valid in situations where the density
stratification is unimportant, a condition that is made more specific in the following chapters, especially in section 5.6. However, analogous shallow-water equations can also be
expressed for systems consisting of discrete, constant-density layers.

## 2.5.    Evaluating Ocean Observations

Regardless of whether one is a seagoing oceanographer or a pure theoretician, it is important to be familiar with ocean observations and to know how to interpret them. This is so
even for those who never analyze data per se, because many practitioners refer to measurements to motivate and to evaluate theory. A fundamental knowledge of data analysis
is important because not all observational literature meets the same contemporary standards of observational design and interpretation. Fortunately, there are a number of
excellent books on the subject (e.g., Thomson and Emery, 2014; Wunsch, 2015) that
discuss central topics.

There is one statistical property that is so universally useful that it deserves mention
here. This is the natural scale (be it time or space) of a process, as defined by the data
themselves. Say that $q(t)$ is a time series that has zero mean and is sampled at time increments of $\Delta t$. It is also assumed that $q$ does not include any trend or substantial perfectly
periodic variability such as a seasonal cycle. (If $q$ does include such a deterministic signal,
it can be removed in practice using a least-squares fit). When considering real oceano-
graphic data (as opposed to a sequence of discrete events such as the coin tosses in introductory statistical textbooks) it is natural to ask how long is the time interval over which
an instantaneous observation is representative; that is, how quickly does a measurement
become obsolete? One way to address this is to compute a correlation of $q$ now with $q$
at some later time:

$$R_q(n\Delta t) \cong \frac{1}{(M-n-1)\{q^2\}}\sum_M q(t)q(t+n\Delta t), \qquad (2.5.1a)$$

where the variance of $q$ is

$$\{q^2\} \cong \frac{1}{M-1}\sum_M q(t)^2, \qquad (2.5.1b)$$

and $n$ is an integer, $M$ is the number of points in the time series, and $R_q$ is called the *autocorrelation function* of $q$. For $n=0$, $R_q(0)=1$, but for very large lag $n\Delta t$, the two time series

$q(t)$ and $q(t+n\Delta t)$ ought to be unrelated, so (multiplying two unrelated time series together, which gives another random series) one expects that $R_q(\infty) = 0$. For example, at a fixed location, this morning's weather conditions might give a useful guess at this afternoon's weather (i.e., there is some *persistence*), but today's weather tells us little about the weather 2 weeks from now.

What is of particular interest is how quickly $R_q$ approaches 0; that is, how strong is the persistence? The more slowly $q$ changes in time, the broader the autocorrelation function (the more slowly it falls toward zero). This information can be crudely summarized by calculating the *independence time scale*:

$$\tau_q = \int_0^\infty R_q(\xi)d\xi. \tag{2.5.2}$$

This quantity[3] can be thought of as a simple measure of how long (over how many time lags) a time series is correlated with itself. Put another way, it is a measure of how long one would need to wait to make another measurement and get a statistically independent sample. If measurements are made over a time separation large relative to $\tau_q$, (i.e., if $\Delta t/\tau_q \gg 1$), then there is an interval between observations where there is no useful information about $q$: gappy, relatively infrequent sampling like this is called *aliased*. It means that each observation is statistically independent, but it also means there is not enough information to plot a time series meaningfully because of the overly large gaps. When data are oversampled (i.e., much more frequently than $\tau_q$, so that $\Delta t/\tau_q \ll 1$, the time scale gives an indication of how often there is a statistically independent data point. That is, the time scale helps in estimating the amount of redundant data.

This discussion assumes that $q$ varies in time, but the ideas hold just as well if the sampling and lags are in space instead. For example, an independence length scale $\Lambda_q$ is the integral of the autocorrelation (analogous to 2.5.2), but as a function of spatial (rather than temporal) lag. Ideally, if one knew the independence time and space scales of some ocean process, one could plan optimal sampling that was neither unnecessarily frequent nor aliased, that is with $\Delta t/\tau_q \approx 1$ for time sampling and with $\Delta x/\Lambda_q \approx 1$ in the $x$ direction. Likewise, knowing the space and time scales of some phenomenon can alert the reader as to how to evaluate published analyses, for example, in terms of judging sampling adequacy or statistical confidence.

Additionally, the time scale is useful for evaluating whether a data set can be taken as *synoptic*. Synopticity simply means that a set of measurements was completed quickly enough that the system did not change significantly during the sampling. Thus, a temperature section (for example) can be considered a "snapshot" if the measurements were completed in a time less than about $\tau_q$, but caution is called for if the section took longer than $\tau_q$ to complete. Consequently, the observer needs to plan carefully in order to balance the competing needs for synopticity, spatial resolution, and spatial extent.

---

3 As a practical matter, the upper bound of the integral is typically taken to be about half or less than half the length of an actual finite-length time series. This is because, at large lags $\xi$, the time series fragments involved become shorter, and the noisiness of the calculation increases.

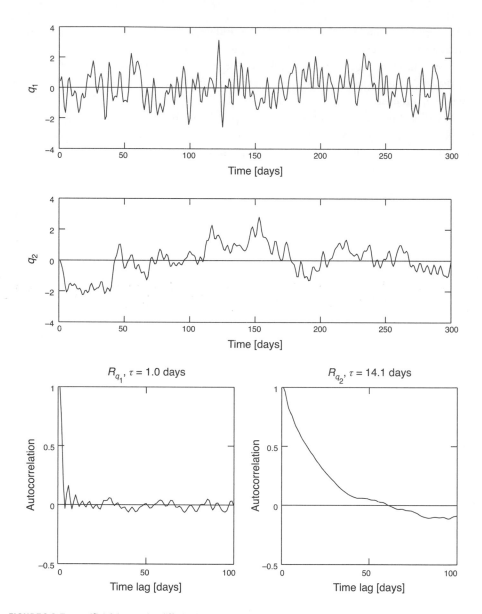

**FIGURE 2.2.** Two artificial time series differing by their inherent time scales (upper two panels), and their respective autocorrelation functions (lower two panels) with independence time scales of 1.0 and 14.1 days, respectively. A comparison of the upper and lower panels demonstrates how the independence time scale characterizes how rapidly the time series fluctuate.

The independence time scale provides a measure of how much information exists in a given time series of length $t_M$ (in days, for example). Specifically, the number of degrees of freedom (independent pieces of information, or number of events) in a time series, for the purpose of evaluating the statistical error about an estimated average, is $t_M/(2\tau_q)$ (see

Thomson and Emery, 2014). Similarly, calculating an analogous independence scale is also required for estimating confidence in a correlation between two time series.

Some of the preceding points are illustrated by Figure 2.2. In the upper two panels are two artificial time series, where $q_1(t)$ has relatively more high-frequency variability than does $q_2(t)$. One way to see that is to compare how frequently $q_1$ crosses zero relative to $q_2$, which has long intervals (up to about 50 days) with no sign changes. Thus, the natural time scale in $q_1$ is shorter than that for $q_2$. This conclusion, of course, is borne out by the autocorrelation functions (lower two plots): the function for $q_1$ approaches zero (as a function of time lag) far more quickly than does the curve for $q_2$. This difference is summarized neatly by the independence time scale $\tau_q$ values, which are 1.0 day for the rapidly varying $q_1$, and 14.1 days for $q_2$. These time scales, in turn, mean that for the purposes of evaluating a mean, $q_1$ has about 14 times as many degrees of freedom than does $q_2$: slowly varying time series of the same duration $t_M$ contain less information (i.e., fewer events or realizations) than rapidly varying ones.

Again, this issue of how much information is present in a time series does not arise in situations where every observation, like every toss of a coin, is known to represent an independent event. Many introductory statistical texts deal only with discrete, independent events of that sort. The independence question becomes important when sampling a system, such as the ocean or atmosphere, where there is often some persistence from one sampling time to the next.

# 3
# Boundary Layers

## 3.1.   Introduction: Where the Ocean Meets the Atmosphere and the Bottom

The contact zones between the sea and what lies above and below are the usually turbulent boundary layers, typically 5–40 m thick, that occupy the uppermost and the lowermost parts of the water column (Figure 3.1). In this example, the boundary layers stand out because their cross-shelf flow opposes that in the interior (i.e., at mid-depth). The near-surface zone absorbs the wind stress and the immediate effects of heating and cooling. How this boundary layer interacts with the remainder of the water column then structures the ocean's response to the wind. At depth, the bottom stress affects circulation both in the boundary layer and, less directly, throughout the water column. These turbulent zones are found nearly everywhere in the ocean, but they are especially important in continental shelf waters, where boundary layers might occupy 50% or more of a 100 m water column, versus perhaps 1% in a 4000 m deep ocean.

Oceanic boundary layers are, by their nature, highly time-dependent features. For example, the thickness of the surface boundary layer depends on the interplay of stabilizing (surface heating, lateral density advection) and destabilizing (surface cooling, lateral advection, turbulence generation by wind stress and wave breaking) influences, all of which vary on time scales of hours or longer. The level of turbulence is determined by the competition of these many factors, and it is often parameterized in terms of an eddy viscosity $A$, which in practice can involve complicated formulations (e.g., Wijesekera et al., 2003). When turbulent processes dominate in the upper ocean, temperature and salinity there are homogenized, and a distinct surface well-mixed layer is formed. Similarly, the thickness of the bottom boundary layer depends on ambient stratification, lateral advection, and bottom stress, again causing variability on similar scales. Within and adjoining these active layers, there are localized currents responding to the stresses. While the interplay of all these processes is indeed complex, it is also true that some important simplifications are remarkably useful.

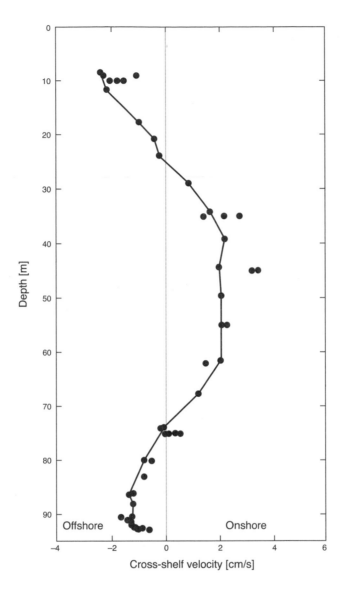

**FIGURE 3.1.** An example of boundary layer structures over the shelf: a vertical profile of seasonal mean (fall and winter: seasons not dominated by coastal upwelling) cross-shelf currents from midshelf off California, north of San Francisco. In this case, flow in the boundary layers opposes that in the interior of the water column: the interior region has positive (onshore) flow, and the surface and bottom boundary layers are identified with offshore flow. Adapted from Lentz and Trowbridge (2001).

The formulation of a boundary layer process needs explanation. One approach is to consider, for example, the hydrostatic momentum equations, averaged over turbulent fluctuations (see chapter 2.3)

$$u_t + uu_x + vu_y + wu_z - fv = -\frac{1}{\rho_0}p_x + \frac{1}{\rho_0}(\tau^x)_z, \tag{3.1.1a}$$

$$v_t + uv_x + vv_y + wv_z + fu = -\frac{1}{\rho_0}p_y + \frac{1}{\rho_0}(\tau^y)_z, \tag{3.1.1b}$$

$$0 = -p_z - g\rho_V. \tag{3.1.1c}$$

Subscripted independent variables represent partial differentiation, $(u, v, w)$ are the velocity components in the $(x, y, z)$ directions, $t$ is time, $f$ is the Coriolis parameter, $\rho_0$ is a constant reference density, $\rho_V$ is the variable part of density, $(\tau^x, \tau^y)$ is the turbulent stress vector (e.g., 2.3.4a), and $g$ is the acceleration due to gravity. When it is possible to linearize the equations of motion (i.e., with a small Rossby number, $\hat{u}/(fL) \ll 1$, where $\hat{u}$ is a typical velocity, and $L$ is a representative horizontal length scale), it is common practice to divide the velocity field into an interior component of flow $(u_I, v_I)$ associated with the pressure field

$$u_{It} - fv_I = -\frac{1}{\rho_0}p_x, \tag{3.1.2a}$$

$$v_{It} + fu_I = -\frac{1}{\rho_0}p_y, \tag{3.1.2b}$$

$$0 = -p_z - g\rho_V, \tag{3.1.2c}$$

and a boundary layer component $(u_E, v_E)$ associated with the vertical stress gradients:

$$u_{Et} - fv_E = \frac{1}{\rho_0}(\tau^x)_z, \tag{3.1.3a}$$

$$v_{Et} + fu_E = \frac{1}{\rho_0}(\tau^y)_z. \tag{3.1.3b}$$

The total velocity is then $u = u_I + u_E$, and so on. It is often convenient to represent the turbulent stresses in terms of an eddy viscosity $A$:

$$(\tau^x, \tau^y) = \rho_0 A(u_{Ez}, v_{Ez}). \tag{3.1.3c}$$

Thus, the boundary layer $(u_E, v_E)$ contribution has been separated out as a relatively simple, one-dimensional problem (3.1.3), and its solution is allowed to be a slowly varying function of $x$ and $y$. If the nonlinear terms in (3.1.1) are not negligible, then this separation into interior and boundary layer components becomes less straightforward.

This chapter deals only with the boundary layer problem (3.1.3) in the northern hemisphere, and it occasionally includes some advective effects as well. Later chapters then apply these results to a wide range of coastal ocean phenomena.

## 3.2.    Some Simple Results: Surface Boundary Layer

The classical boundary layer problem on a rotating planet (Ekman, 1905) assumes that the eddy viscosity $A$ is a constant. In this case, the steady boundary layer problem (3.1.3) becomes

$$-fv_E = Au_{Ezz},$$

(3.2.1a)

$$fu_E = Av_{Ezz},$$

(3.2.1b)

which is simplified by defining $\mu_E = u_E + iv_E$. Then (3.2.1) becomes

$$if\mu_E = A\mu_{Ezz}.$$

(3.2.2)

Solutions to this have the form $\exp(\Gamma z)$, where

$$if = A\Gamma^2,$$

(3.2.3)

so that

$$\Gamma = \pm\sqrt{\frac{f}{2A}}(1+i) = \pm(1+i)/\delta_E,$$

(3.2.4)

that is, one root corresponds to downward decay, and one to upward decay. Note that the scale thickness, $\delta_E = \sqrt{2A/f}$, remains finite as long as $f$ does not vanish; that is, rotation implies a limited boundary layer thickness even if turbulence (eddy viscosity) does not vary with depth.[1]

At this point, it is useful to assume that the water depth $h$ is large relative to the boundary layer scale; that is, $h/\delta_E = h\sqrt{f/(2A)} \gg 1$ (but see section 6.2 for when this is not true). In this deep-water case, the surface and bottom boundaries can be treated separately. At the ocean surface ($z=0$), it is normal to assume that the wind stress

$$T_0 = \tau_0^x + i\tau_0^y$$

(3.2.5)

is given. Only the positive, downward-decaying root of (3.2.4) is appropriate near the surface, so

$$\mu_E = C \exp\left[\sqrt{\frac{f}{2A}}(1+i)z\right].$$

(3.2.6)

The surface, stress-matching boundary condition is

$$T_0 = \rho_0 A\mu_{Ez} = \rho_0 C\sqrt{\frac{fA}{2}}(1+i) \qquad \text{at } z = 0,$$

(3.2.7)

so that

$$C = \frac{1}{\rho_0}\sqrt{\frac{1}{2Af}}\left[(\tau_0^x + \tau_0^y) + i(-\tau_0^x + \tau_0^y)\right].$$

(3.2.8)

---

1 This result can be compared with the oscillating [$\exp(i\omega t)$, where $\omega$ is the frequency] but nonrotating (from 3.1.3)

$$i\omega u_E = Au_{Ezz}$$

or, using $u_E = \exp(\gamma z)$,

$$\gamma = \pm\sqrt{\frac{\omega}{2A}}(1+i).$$

In this case, the boundary layer depth is kept finite because the oscillating stress takes time to penetrate over depth and is constantly reversing. In this fluctuating case, however, there is no veering in the velocity.

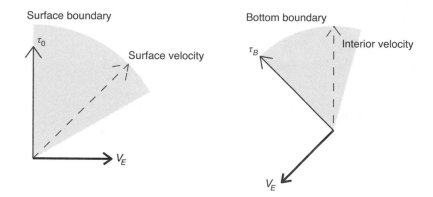

**FIGURE 3.2.** Diagram showing the directional offsets of boundary layer currents, stress $\tau$, and transport $V_E$ for surface (left) and bottom (right) boundary layers in the northern hemisphere. The boundary layer component of surface velocity $v_S$ and the interior velocity above the bottom boundary layer $v_I$ both appear as dashed arrows. The 45° velocity offsets are peculiar to the constant eddy viscosity problem, so the shading around $v_S$ and $v_I$ is meant to indicate that their actual directions vary depending on the eddy viscosity parameterization; however, transport is always orthogonal to the stress in the steady, linear problem.

This result can then be used to find solutions for the boundary layer flow $(u_E, v_E)$. For simplicity, assume that $\tau_0^x = 0$, so that

$$u_E = \frac{\tau_0^y}{\rho_0}\sqrt{\frac{1}{2Af}}\exp\left(\sqrt{\frac{f}{2A}}z\right)\left[\cos\left(\sqrt{\frac{f}{2A}}z\right) - \sin\left(\sqrt{\frac{f}{2A}}z\right)\right], \tag{3.2.9a}$$

$$v_E = \frac{\tau_0^y}{\rho_0}\sqrt{\frac{1}{2Af}}\exp\left(\sqrt{\frac{f}{2A}}z\right)\left[\cos\left(\sqrt{\frac{f}{2A}}z\right) + \sin\left(\sqrt{\frac{f}{2A}}z\right)\right]. \tag{3.2.9b}$$

The solution describes a spiral, decaying with depth, starting with a surface velocity which is 45° to the right (in the northern hemisphere) of the wind stress. Further, the result can be integrated in $z$ from $-\infty$ to 0 to obtain an expression for the transport

$$(U_E, V_E) = \left(\frac{\tau_0^y}{\rho_0 f}, 0\right). \tag{3.2.10}$$

The relative directions of stress, surface velocity, and transport, for the northern hemisphere, are illustrated in the left panel of Figure 3.2. Because the angle of the surface velocity relative to the stress depends on the particular eddy viscosity form, this velocity vector is surrounded by shading, as a reminder of this model-to-model variability.

The solution (3.2.9) illustrates the well-known Ekman layer, a boundary layer characteristic of rotating systems. Using a constant eddy viscosity $A$ is actually a rather unrealistic assumption, but it allows simple solutions, and it makes the point that the boundary layer thickness is limited by rotation even when active turbulence (eddy viscosity) extends to great depths. Thus, the detailed structure (3.2.9), including the 45° offset of surface velocity, is an idealization, although the general spiral character does appear to be robust in the real ocean (e.g., Chereskin, 1995). On the other hand, the expression for surface Ekman transport (3.2.10) is very robust, since it can be obtained by vertically integrating (3.2.1) from some depth where the turbulent stress vanishes (i.e., where $Au_{Ez} = Av_{Ez} = 0$) to the surface. Thus, provided that the flow is steady and linear,

$$(U_E, V_E) = \frac{1}{\rho_0 f}(\tau_0^y, -\tau_0^x) \tag{3.2.11}$$

is a general, yet simple, result that is independent of the eddy viscosity form.

## 3.3.   Some Simple Results: Bottom Boundary Layer

The near-bottom region needs to be considered separately, because a different boundary condition leads to contrasting results. Specifically, an appropriate condition when the eddy viscosity is constant is that the total velocity vanishes at the bottom ($z = -h$); that is,

$$u_I + u_E = 0, \; v_I + v_E = 0 \qquad\qquad \text{at } z = -h. \tag{3.3.1 a, b}$$

Using the negative (upward-decaying) root in (3.2.4), the solution is

$$\mu_E = -(u_I + iv_I)\exp\left[-\sqrt{\frac{f}{2A}}(i+1)(z+h)\right], \tag{3.3.2}$$

an Ekman spiral that decays upward. The bottom stress (in the northern hemisphere) is then

$$\tau_B^x = \rho_0\sqrt{\frac{fA}{2}}(u_I - v_I), \tag{3.3.3a}$$

$$\tau_B^y = \rho_0\sqrt{\frac{fA}{2}}(u_I + v_I); \tag{3.3.3b}$$

that is, the stress is at an angle relative to the interior velocity (Figure 3.2, right panel), although the actual angular offset depends on the eddy viscosity model that is applied. The bottom Ekman transport, however, remains orthogonal to the bottom stress (by the same logic that yields 3.2.11) when flow is steady and linear, so that

$$U_E = -\frac{\tau_B^y}{\rho_0 f} = -\sqrt{\frac{A}{2f}}(u_I + v_I), \tag{3.3.4a}$$

$$V_E = \frac{\tau_B^x}{\rho_0 f} = \sqrt{\frac{A}{2f}}(u_I - v_I), \tag{3.3.4b}$$

where the first equality in each case is a general result for steady linear flow but the second equality applies *only* with a constant eddy viscosity.

For the bottom boundary layer, it can be difficult to relate the interior velocity to the bottom stress in realistic conditions with variable eddy viscosity. This complication occurs because, in practice, the eddy viscosity itself depends on the velocity, making the problem inherently nonlinear in practice. Some aspects of the problem are quite general, however, such as the relation of stress to transport, but other aspects (such as the relation of bottom stress to interior velocity) are less universal.

## 3.4.   Ekman Compatibility Condition

Boundary layer transports are the pathways by which the effects of surface and bottom stresses are ultimately conveyed to the interior water column (i.e., to the part of the water column that does not contain turbulent boundary layers). In an idealized situation, Ekman transport affects the interior in two ways. One involves an idealized impermeable lateral boundary, for example, for a coastal wall at $x = 0$,

$$0 = \int_{-h}^{0} u \ dz = \int_{-h}^{0} u_I dz + U_E, \tag{3.4.1}$$

where $U_E$ is taken to include both a surface Ekman transport (due to wind forcing) and a bottom component (due to friction). This condition implies an unresolved vertical transport connecting $u_I$ and $U_E$ at or inshore of the imposed boundary (see chapter 6 for how this happens). But the important point about condition (3.4.1) is that the boundary layer and interior flows are strongly coupled at any inshore barrier, and this connection is often the most effective means for conveying forcing and dissipation to the interior from the boundary layers.

Away from lateral boundaries, the Ekman layers can exchange water with the interior if the boundary stresses vary horizontally. To demonstrate this second mechanism, consider that the continuity equation applies to the interior and Ekman flow components separately, so that

$$0 = u_{Ex} + v_{Ey} + w_{Ez}. \tag{3.4.2}$$

For the surface case, this is vertically integrated from a depth where stress just vanishes, $z = -\eta_0$, to the surface to obtain

$$0 = w_{top} - w_{bot} + U_{Ex} + V_{Ey}, \tag{3.4.3}$$

where $w_{top}$ is the vertical velocity at the free surface (where $w \approx 0$), and $w_{bot}$ is $w_E$ just below the turbulent layer, at $z = -\eta_0$. Using the general expressions for steady, linear Ekman transport (3.2.11) yields a statement that the vertical velocity is determined by convergence/divergence in the surface boundary layer:

$$w_{bot} = w_E = \frac{1}{\rho_0 f}(\tau_{0x}^y - \tau_{0y}^x). \tag{3.4.4}$$

This sets the vertical velocity at the upper boundary of the inviscid interior, thus forging a link between the boundary layer and interior flow fields. Note that this result does not depend on the eddy viscosity formulation. Physically, the surface boundary layer is either ejecting water downward into the interior (Ekman pumping) or pulling it upward into the near-surface region (Ekman suction).

Using a similar approach near a flat bottom, the vertical velocity at the top of the bottom boundary layer ($z = -h + \eta_0$) is

$$w_{top} = w_E = \frac{1}{\rho_0 f}(\tau^y_{Bx} - \tau^x_{By}).$$

$$(3.4.5)$$

While this expression holds true for steady, linear flow and any eddy viscosity formulation, the relation of stress to interior velocity is sensitive to the closure scheme. Revisiting the constant $A$ velocity-stress relation (3.3.3) but allowing for a more general, but still linear, dependence of stress on interior velocity (e.g., with a different $A$ parameterization or bottom boundary condition), leads to

$$\tau^x_B = \rho_0 \sigma_F(u_I - \gamma v_I),$$

$$(3.4.6a)$$

$$\tau^y_B = \rho_0 \sigma_F(\gamma u_I + v_I),$$

$$(3.4.6b)$$

where $\sigma_F$ and $\gamma$ are model-dependent parameters, and $\gamma$ is expected to be $O(1)$ or smaller. Thus, the veering in near-bottom velocity (accounted for in $\gamma$) does not have to be $45°$ relative to the stress; this range of possibilities is expressed schematically by the gray shading in Figure 3.2. Using the form (3.4.6) in (3.4.5) leads to

$$w_E = \frac{\sigma_F}{\rho_0 f}[(v_{Ix} - u_{Iy}) + \gamma(u_{Ix} + v_{Iy})].$$

$$(3.4.7)$$

Very often, on time scales longer than the inertial (and the present derivation requires this to be true in order to neglect time derivatives in 3.1.3), the divergence term multiplied by $\gamma$ is small compared with the relative vorticity $v_{Ix} - u_{Iy}$. For example, the $\gamma$ term is identically zero if the interior flow is geostrophic and $f$ is a constant. In any case, neglecting the $\gamma$ term in (3.4.7) is consistent with using the simple stress form

$$(\tau^x_B, \tau^y_B) \cong \rho_0 \sigma_F(u_I, v_I),$$

$$(3.4.8)$$

which is often quite reasonable if the detailed boundary layer flow is not of interest.

If the interior velocity has a substantial (in the Rossby number sense) lateral shear, the consequent momentum advection in (3.1.1) can modify the boundary layer flow (Stern, 1965; Niiler, 1969). Consider the surface boundary layer, and say there is a steady mean interior velocity $(0, v_0(x))$ independent of $z$. The steady Ekman problem (from 3.1.1) becomes

$$-fv = (Au_z)_z,$$

$$(3.4.9a)$$

$$uv_{0x} + fu = (Av_z)_z.$$

$$(3.4.9b)$$

Vertically integrating these over the boundary layer then leads to

$$U_E = \frac{\tau^y_0}{\rho_0(f + v_{0x})},$$

$$(3.4.10a)$$

$$V_E = -\frac{\tau^x_0}{\rho_0 f},$$

$$(3.4.10b)$$

so that (3.4.3) becomes (assuming no $y$ variations)

$$w_{bot} = w_E = U_{Ex} = \left[ \frac{\tau_0^y}{\rho_0(f + v_{0x})} \right]_x. \tag{3.4.11}$$

Thus, a spatially uniform wind stress can give rise to upwelling or downwelling at the base of the boundary layer if the ambient velocity has a nonzero gradient in relative vorticity, that is, if $v_{0xx} \neq 0$. The problem becomes considerably more complicated if the ambient flow is more complex—for example, if it includes curvature—but the underlying physics remains relevant (e.g., Wenegrat and Thomas, 2017). Indeed, there is observational verification of this vorticity-induced vertical velocity in the upper ocean (e.g., Gaube et al., 2015).

Nonlinearity associated with strong interior shears affects the bottom boundary layer as well, with some differences (e.g., Brink, 1997). For one, if the ambient flow contains relative vorticity, there will already be a boundary layer vertical velocity (from 3.4.7), even for a small Rossby number. In other words, because the bottom stress depends on the interior velocity, nonlinearity is not *required* to generate a vertical velocity (as is true near the surface with a constant wind stress), even if it does cause modifications. Further, within the boundary layer, the total velocity weakens near the bottom, so that even if the local Rossby number $|v_x|/f$ is substantial and depth-independent in the interior, it weakens near the boundary. In contrast, the surface boundary layer problem does not require the total velocity to weaken near the boundary, so $v_0$ can often be taken as vertically constant in (3.4.9).

## 3.5.   The Surface Mixed Layer

The laminar Ekman layer models of section 3.2 are, of course, extreme idealizations, because turbulence in the ocean is far from constant in either space or time. Specifically, density stratification inhibits turbulence, so that far from the surface processes that generate shears, the water column may become relatively laminar. Expressed another way, the eddy viscosity $A$ is expected to be much larger near the surface than at depth (see exercise A.2 in the appendix). Thus, a very useful approach to the upper ocean is to consider a surface mixed layer where waters are homogenized by the turbulence owing to such effects as wind stress, breaking waves, surface cooling, and velocity shears (e.g., D'Asaro, 2014), all of which are highly time-dependent in reality. This variability, in turn, means that the homogenized upper ocean is also transitory, disappearing when there is sufficient surface warming (which stabilizes the stratification), or deepening when there is strong enough turbulence. For example, on the Peruvian shelf (Figure 3.3), wind stress varies on diurnal (sea breeze) and event time scales. Further, waters are warmed during the daytime and cooled at night. As a result, the mixed-layer depth varies on time scales of hours and longer, including a clear diurnal cycle.

A common idealization is to treat the mixed layer as a slab having vertically constant density and horizontal velocity, and overlying a stratified, motionless interior (Figure 3.4).

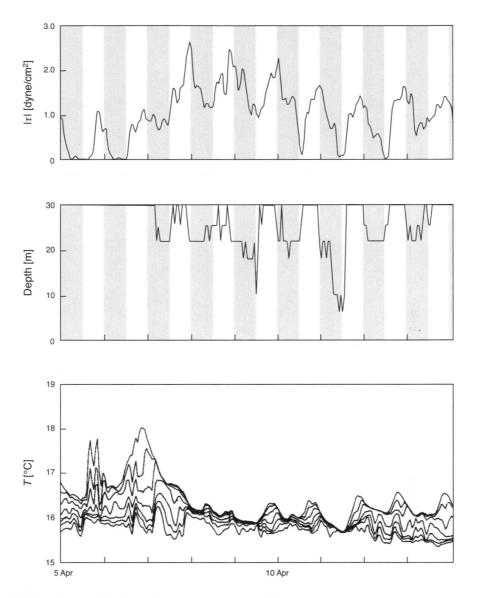

**FIGURE 3.3.** Time variability of the surface boundary layer in a stratified ocean: measurements from the upper ocean at midshelf off Peru near 15°S, obtained from a fixed mooring in 1977. Upper panel: magnitude of the surface wind stress (dyne/cm²). Middle panel: mixed-layer depth, defined as the depth over which temperature agrees with the shallowest measurement to within 0.02°. The shaded areas denote nighttime, and the areas in between denote daytime. Lowest panel: time series of temperature at depths of 2.1, 4.6, 8.1, 12, 16, 20, 24 m. Waters are well mixed when different lines overlie exactly. Adapted from Brink et al. (1983).

This structure is consistent with having the eddy viscosity $A$ be large (in the sense of $A/(fh_{ML}^2) \gg 1$, where $h_{ML}$ is the mixed-layer depth) and constant within the mixed layer, and zero below. One advantage of this approach is that, however simply, it begins to address the interplay of mixing and water column stratification. Several comprehensive

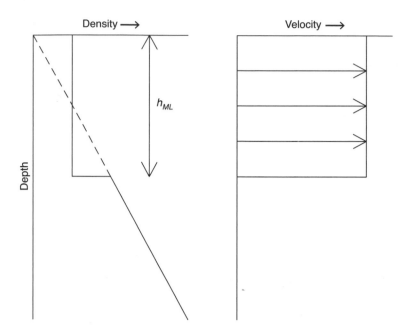

**FIGURE 3.4.** Schematic of an idealized surface mixed layer having thickness $h_{ML}$. Left panel: density, with the density profile before mixing shown as a dashed line. Right panel: velocity.

idealized models of the mixed layer exist and account for heating, cooling, wind forcing, and so on (e.g., Price, et al., 1986). However, for simplicity, the approach here is to treat one effect at a time.

First, consider mixed-layer deepening due to turbulence generated by shear at the base of the mixed layer (Pollard et al., 1973). Assume that the initial density profile is simply

$$\rho = \rho_0 + \gamma z, \tag{3.5.1}$$

where $\gamma = \rho_z < 0$ is a constant, and that there is no surface heating or cooling. In that case, the density of a mixed layer of thickness $h_{ML}$ is the average of the initial density over that depth (Figure 3.4), or

$$\rho_{ML} = \rho_0 - \gamma h_{ML}/2, \tag{3.5.2}$$

and the density jump across the bottom of the mixed layer is

$$\Delta \rho = -\gamma h_{ML}/2. \tag{3.5.3}$$

Assuming the wind is only in the $y$ direction and there is no horizontal variability, the layer-integrated momentum equations are

$$U_t - fV = 0, \tag{3.5.4a}$$

$$V_t + fU = \frac{1}{\rho_0} \tau_0^y, \tag{3.5.4b}$$

where $(U, V) = h_{ML}(u, v)$ because of the slab-like flow (Figure 3.4). Now, assume the wind stress is given by

$$\tau_0^y = 0 \qquad \text{for } t < 0, \tag{3.5.5a}$$

$$\tau_0^y = \tau_A \qquad \text{for } t > 0, \tag{3.5.5b}$$

and the ocean is initially at rest. Then, for $t > 0$,

$$U = h_{ML}u = \frac{\tau_A}{f\rho_0}[1 - \cos(ft)], \tag{3.5.6a}$$

$$V = h_{ML}v = \frac{\tau_A}{f\rho_0}\sin(ft); \tag{3.5.6b}$$

that is, the solution consists of the steady Ekman transport plus an inertial oscillation (frequency $f$). In a realistic context, the inertial oscillation would eventually decay away owing to propagation or frictional dissipation. Note, too, that the result for transports (3.5.6) is fairly general, since (3.5.4) can be obtained by first vertically integrating over a boundary layer (as in section 3.4) without assuming a slab-like structure.

At this point, the solution is incomplete because $h_{ML}$ remains undefined. The problem is closed by introducing a bulk Richardson number (equivalent to an inverse Froude number)

$$R_B = \frac{g\Delta\rho h_{ML}}{\rho_0(u^2 + v^2)}, \tag{3.5.7}$$

that is, the ratio of the squared speed of an interfacial internal gravity wave ($g\Delta\rho h_{ML}/\rho_0$) to the squared velocity jump at the bottom of the layer. Physically, this is the ratio of the stabilizing tendency of stratification versus the destabilizing effect of a velocity shear. If $R_B$ is less than some critical value $R_{BC}$, then turbulence is generated at the interface, and the layer deepens until $R_B$ increases to the critical value. As the layer depth $h_{ML}$ increases, the density jump $\Delta\rho$ (3.5.3) increases, while the velocity jump decreases. Both of these effects tend to increase (stabilize) $R_B$. Using (3.5.3) and (3.5.6), and assuming that the mixed-layer thickness is always critical or greater, leads to

$$R_{BC} \leq \frac{\rho_0 g\Delta\rho h_{ML}^3 f^2}{2\tau_A^2[1 - \cos(ft)]}. \tag{3.5.8}$$

Initially, the quantity in brackets increases until $t = \pi/f$, when it reaches a maximum of 2, which recurs every $2\pi/f$. To honor the critical condition, $h_{ML}$ must also increase until $t = \pi/f$. After that time, the velocity jump at the layer bottom never gets larger, and so $h_{ML}$ remains at the $t = \pi/f$ value. Thus, with the maximum shear and $h_{ML}$,

$$R_{BC} = \frac{f^2 N^2 h_{ML}^4}{8u_*^4}, \tag{3.5.9a}$$

where the buoyancy frequency squared is based on the initial stratification

$$N^2 = \frac{-g\rho_z}{\rho_0} = \frac{-g\gamma}{\rho_0}, \tag{3.5.9b}$$

and the *friction velocity* is

$$u_* = \left(\frac{|\tau_A|}{\rho_0}\right)^{1/2}. \tag{3.5.9c}$$

Rearranging (3.5.9a) leads to

$$h_{ML} = (8R_{BC})^{1/4}\frac{u_*}{\sqrt{fN}} = G\frac{u_*}{\sqrt{fN}}, \tag{3.5.10}$$

which is often called the Pollard-Rhines-Thompson (1973) depth. The coefficient $G$ has been reported in the literature to have values ranging from around 0.6 to 1.7, and the formulation (3.5.10) often agrees well with midshelf mixed-layer observations in several regions of coastal upwelling (e.g., Lentz, 1992).

Surface heat fluxes modify the problem. Warming produces lighter water, which increases the density contrast $\Delta\rho$ at the bottom of the mixed layer, and so, from (3.5.8), the mixed layer is shallower than in the absence of warming. Similarly, surface cooling leads to a deeper mixed layer. In fact, sustained cooling causes mixing even in the absence of a wind stress owing to gravitational instability (heavy water appearing above lighter) and can eventually cause the entire water column to become well mixed. The good agreement that Lentz (1992) obtained between measurements and (3.5.10) is actually somewhat surprising, because he found that time series of mixed-layer depth are not correlated with surface heat flux, as might be expected. His explanation was that the destabilizing cooling effect of transporting cool upwelled water offshore in the surface boundary layer appears to balance the stabilizing effect of surface warming. This, in turn, suggests that the simple Pollard-Rhines-Thompson formula (3.5.10) may not apply as well in other coastal settings where wind-driven upwelling is not so dominant.

Other illustrative limiting behaviors for surface mixing can be found for steady winds after the depth (3.5.10) is reached. One class of model for this phase is expressed in terms of the gravitational potential energy balance of the mixed layer (Niiler, 1975). Specifically,

$$(h_{ML})_t\left[N^2\frac{h_{ML}^4}{2} + \dots\right] = 2m_0u_*^3h_{ML}^2 - \frac{\alpha_T g}{C_p\rho_0^2}F_T h_{ML}^3, \tag{3.5.11}$$

where the constant $m_0$ reflects the efficiency of the turbulence generated by the wind stress at the ocean's surface for deepening the mixed layer, $F_T$ is a surface heat flux (positive for warming), $\alpha_T$ relates temperature changes to density changes, and $C_p$ is the heat capacity of water. The $N^2$ term on the left-hand side reflects changes in the gravitational potential energy, and other terms in the brackets have been omitted for simplicity. Altogether, (3.5.11) is a statement that the gravitational potential energy changes are due to (1) deepening associated with turbulence input directly from the wind stress (the $m_0$ term: a source of turbulence distinct from that due to shear at the bottom of the mixed layer) and to (2) changes in mixed-layer density associated with surface heating or cooling (the $F_T$ term).

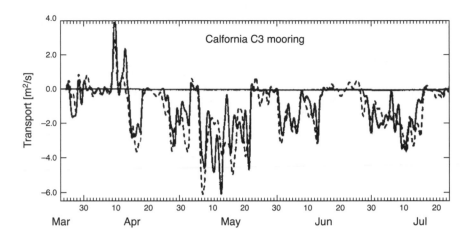

**FIGURE 3.5.** Observed time variability in surface boundary layer transport: results from a midshelf mooring during the upwelling season north of San Francisco. Dashed line: theoretical Ekman transport, $U_E = \tau_0^y / (\rho_o f)$. Solid line: the integral of observed cross-shelf flow over a depth of 1.5 times the observed mixed-layer depth. Current measurements are from five depths in the upper 30 m of the water column. Records have been smoothed to eliminate tides, the sea breeze, and internal waves. Adapted from Lentz (1992).

If $F_T = 0$, solving for $h_{ML}$ in (3.5.11) results in an expression for slow mixed-layer deepening,

$$h_{ML} = (4m_0)^{1/3} \frac{u_*}{N} (tN)^{1/3}, \tag{3.5.12}$$

sometimes called the Krauss-Turner regime. Typically, this slow erosion would be detectable with sustained winds after the initial depth (3.5.10) was reached in half an inertial period.

Balancing the two right-hand-side terms in (3.5.11) shows that there can be a steady-state mixed-layer depth

$$h_{ML} = \frac{2m_0 u_*^3 C_p \rho_0^2}{\alpha_T g F_T}. \tag{3.5.13}$$

This reflects a balance of surface warming (tending to shoal the mixed layer) and the mixing caused by a sustained surface wind stress. All these asymptotic states (3.5.10 and especially 3.5.12 and 3.5.13), of course, are somewhat artificial because they assume a steady, sustained surface forcing, but they do illustrate some of the mechanisms at play.

What does the flow in an actual mixed layer look like? Lentz (1992) examined detailed moored measurements from three regions where coastal upwelling is active and winds are primarily alongshore. He found that, indeed, the instantaneous cross-shelf flow in the mixed layer is slab-like, but not all the Ekman transport occurs in the mixed layer (Figure 3.5). Rather, there is a region below the mixed layer, having about half the thickness of the mixed layer itself, where part of the Ekman transport occurs. Within this region the turbulent stress must be nonzero to balance the cross-shelf transport. This,

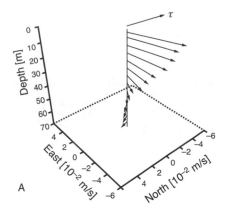
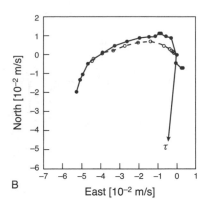

**FIGURE 3.6.** Two views of mean horizontal velocity in a surface boundary layer, demonstrating an Ekman spiral. Left panel: measured mean (over 15 days) wind stress and currents over the upper 70 m of the water column at an open-ocean location. Current at 70 m has been subtracted at all depths to isolate the Ekman flow component. The currents are continuously rotated so they are always consistently oriented relative to the wind stress. Right panel: hodograph plot of mean currents and wind stress (arrow) from the same data set. Current measurements are at the dotted locations beginning with 2 m depth (leftmost point) and then at 4 m increments below. The solid line represents observations, and the dashed line is the result of a sophisticated mixed-layer model. Adapted from Price et al. (1986).

in turn, implies active mixing (hence nonzero eddy viscosity) in the transitional region but not so strongly turbulent as to eliminate the stratification. What about the Ekman spiral that appears so neatly with constant eddy viscosity? Even though the flow is slab-like instantaneously, the layer thickness is continually changing, so that, averaged over many days, a spiral pattern can still emerge (e.g., Figure 3.6).

## 3.6.    The Bottom Boundary Layer with a Flat Bottom

The turbulent layer near the bottom can consist of different overlapping regions, each with a different balance of forces. The zones close to the ocean bottom (within a few meters) are critical for setting the stress, given current and wave conditions higher up. The $O(10\text{ m})$-thick outer regions of the boundary layer are where rotation is important, the net transport perpendicular to the stress occurs, and the stress direction is determined. In reality, these regions blend smoothly from one to the other, but they are all important for the overarching goal of determining the bottom stress corresponding to a given interior current. A good beginning is to isolate and treat the fundamental inner layer: the constant-stress or *logarithmic layer*.

The starting assumptions are that the flow is turbulent, the water column is unstratified, and the system is not rotating (or that interest is confined close enough to the bottom for rotation not to matter[2]). Scaling arguments, applied to a horizontal momentum

---

2 Specifically, scaling (3.1.3) and using the result (3.6.8) shows that neglecting rotation requires that $zf/(\kappa u_*) \ll 1$ (where $\kappa$ is von Kármán's constant). Also, neglecting time dependence requires that $z\omega/(\kappa u_*) \ll 1$,

equation such as (3.1.1a), lead to the conclusion that near the boundary, the turbulent stress is independent of height within the boundary layer; that is,

$$\tau_z = 0, \tag{3.6.1}$$

where the stress and the flow (averaged over boundary layer turbulence) are in the same direction. The next question regards the velocity structure in the boundary layer. A standard approach to this question is to ask, what are the dimensional variables in this problem ($\tau$, $\rho_0$, $u$, and $z$, where the bottom is now at $z = 0$) and then to ask, how can they be combined to provide a dimensionally correct expression for the flow profile. First note that $\tau$ and $\rho_0$ can be combined into an expression with the dimensions of velocity:

$$\left( \frac{|\tau|}{\rho_0} \right)^{1/2} = u_*, \tag{3.6.2}$$

where $u_*$ is again the friction velocity. Because the turbulent stress can be written as

$$\tau = -\rho_0 <u'w'>, \tag{3.6.3}$$

(where the angle brackets represent an average over turbulent variables, and a prime represents a turbulent fluctuation), and $u_* = \sqrt{\langle u'w' \rangle}$ can be thought of as representative of the magnitude of the boundary layer turbulent velocity, as is indeed observed in the ocean (e.g., Perlin et al., 2005).

Now, seeking an expression for the velocity profile, one might consider the dimensionally correct expression

$$u \overset{?}{=} Cu_*, \tag{3.6.4}$$

where $C$ is some dimensionless constant. This expression is not very satisfactory, because velocity would be a constant throughout the boundary layer. Rather, flow ought to become weaker near the boundary and approach some interior, or at least larger, velocity far from the boundary. Next, one could try

$$\frac{du}{dz} \overset{?}{=} u_* \frac{1}{\kappa z}, \tag{3.6.5}$$

where $\kappa$ turns out to be the dimensionless von Kármán's constant, typically taken to be about 0.40. This expression can be integrated in $z$ to obtain

$$u = \frac{u_*}{\kappa}(\ln z + E) = \frac{u_*}{\kappa}\ln\left( \frac{z}{z_C} \right) \tag{3.6.6}$$

where $E = -\ln(z_C)$ is a constant of integration. This result has some pleasing properties: $u$ becomes smaller as $z$ becomes small, and it does not grow too rapidly for large $z$. Better yet, this velocity profile agrees well with a wide variety of observations: laboratory, ocean,

---

where $\omega$ is a typical frequency of far-field (nonturbulent) velocity fluctuations, and $z$ is the distance from the bottom.

and atmosphere (e.g., Trowbridge and Lentz, 2018). In fact, this "law of the wall" (3.6.6) is universally accepted as a reasonable representation of a turbulent boundary layer under conditions where (3.6.1) valid. Of course, this result fails far from the bottom: velocity cannot increase indefinitely. Rather, limits on $u$ can be due to Earth's rotation, to time dependence, or to density stratification inhibiting turbulence. Typically, in the ocean, the truly constant stress layer is $O(3 \text{ m})$ thick.

This velocity profile needs to be explored further. For one thing, it does not behave well as $z$ goes to zero. Rather, $u = 0$ at $z = z_C > 0$, and other physics must apply between $z = z_C$ and the physical bottom ($z = 0$). It has been found that if the bottom is perfectly smooth, $z_C = v/u_*$, where $v$ is the molecular viscosity. In this situation, there is a laminar sublayer within $z_C$ of the true boundary, and velocity is zero at the physical bottom. Alternatively, if the bottom is rough, $z_C = z_0$ is called the *roughness height*, and it is assumed that there is a turbulent region below $z = z_0$. Roughly, $z_0 \cong d_r/30$, where $d_r$ is the actual physical height of the bottom bumps, although more sophisticated expressions for $z_0$ exist (e.g., Trowbridge and Lentz, 2018). The transition between the smooth and rough sublayer regimes occurs when the Reynolds number $Re = u_* d_r / v$ is $O(1)$.

An important consequence of (3.6.6) is that it leads to a relation between stress and velocity. Specifically, squaring both sides (and using the definition of $u_*$) leads to

$$|\tau| = \frac{\rho_0 u^2 \kappa^2}{\left[\ln\left(\frac{z}{z_C}\right)\right]^2} \equiv \rho_0 u^2 c_D, \tag{3.6.7}$$

where $c_D$ is called the *drag coefficient*.[3] It is important to remember that $c_D$ is a function of both roughness and of height above the bottom. Typical values of $c_D$ in the oceanographic literature are about $1$–$5 \times 10^{-3}$, but in extremely rough settings such as coral reefs or sea ice, it can be orders of magnitude larger (e.g., Lentz et al., 2017).

Several lines of argument can lead to the logarithmic velocity profile; the simple approach here follows that of Turner (1973), who essentially ignored rotation. A more sophisticated approach was used by Grant and Madsen (1986), who arrived at the same point but accounted explicitly for Earth's rotation and thus an outer boundary layer. Finally, a more traditional approach involves mixing length theory (e.g., Holton and

---

3 Expression (3.6.7) gives rise to a handy, albeit crude, rule of thumb. Say that at the surface of the ocean there are logarithmic boundary layers in both the ocean and the atmosphere. Then, matching stress across the interface,

$$\rho_W c_{DW} u_W^2 = \rho_A c_{DA} u_A^2$$

where a subscript $W$ refers to the ocean water, and $A$ to the atmosphere. Then, assuming that $c_{DW} \cong c_{DA}$,

$$\frac{u_W}{u_A} = \sqrt{\frac{\rho_A}{\rho_W}} \cong 3\%.$$

The assumptions here are very suspect, because waves, stratification, rotation, and other effects are ignored (and the evidence for an ocean surface logarithmic layer is spotty), but the estimate of surface currents being about 3% of the wind speed remains popular in some quarters.

Hakim, 2012) and the argument that the size of turbulent eddies increases with distance from the bottom, so that the eddy viscosity profile is

$$A = \kappa u_* z, \qquad (3.6.8)$$

and

$$A\frac{du}{dz} = \frac{\tau}{\rho_0} = u_*^2. \qquad (3.6.9)$$

Integrating (3.6.9) using (3.6.8) leads again to (3.6.6). One particularly convenient aspect of the eddy viscosity approach is that the $A$ profile can be modified to account for the stabilizing effect of density stratification (e.g., Trowbridge and Lentz, 2018). Many turbulence closure models (Wijesekera et al., 2003 summarized several) modify their eddy viscosity estimates near surface and bottom boundaries in a manner at least qualitatively consistent with (3.6.8). This tapering near boundaries yields useful results, although the observational evidence for a free-surface logarithmic layer is not very conclusive; the boundary moves with waves, breaking waves generate mixing, and other processes such as Langmuir turbulence (e.g., Thorpe, 2004) complicate matters.

A particularly interesting variant on this problem arises owing to short (nonhydrostatic) gravity waves on the continental shelf. These waves (section 6.3) decay with distance from the surface on a length scale of $k^{-1}$, where $k$ is the horizontal wavenumber. Thus, a typical swell with period of 12 s has a substantial horizontal velocity component at the ocean bottom in 100 m of water, that is, across much of the shelf. (Higher-frequency, shorter waves do not penetrate as deeply.) The waves have their own bottom sublayer of thickness scale $u_{*W}/\omega$, where $u_{*W}$ is the friction velocity associated with the wave, and $\omega$ is the wave frequency. The wave boundary layer is thus typically 5–30 cm thick (depending on wave amplitude and frequency) and couples to the outer, thicker constant-stress layer. The result is effectively an enhanced bottom roughness $z_{0W}$ due to the enhanced turbulence in the near-bottom wave boundary layer (Grant and Madsen, 1986). This enhancement can be substantial—for example, doubling the bottom stress—but because wave conditions vary with time, it also means that the drag coefficient (3.6.7) varies with time.

The logarithmic layer is, of course, a substantial simplification. For example, Earth's rotation has been ignored to this point. System (3.2.1) can be solved with $A = \kappa u_* z$, and the result is that the lower part of the boundary layer is approximately logarithmic, like (3.6.6), but the upper part of the boundary layer shows a decaying Ekman spiral with a vertical scale of order $u_*/f$ (e.g., Madsen, 1977, for the surface boundary version). Under reasonable shelf conditions, this would give a rather large boundary layer thickness of about 20–50 m or more, which is not surprising given the idealized, ever-increasing eddy viscosity profile. In reality, either suspended sediments (near the bottom) or water column stratification (away from the bottom) can weaken or even prevent turbulence and hence inhibit the eddy viscosity. Thus, a realistic eddy viscosity profile can strongly modify the boundary layer's structure and constrain its thickness (Grant and Madsen, 1986; Trowbridge and Lentz, 2018). For these reasons, there can be substantial deviations from a logarithmic profile even within a few meters of the bottom, and there are many reasons for the drag coefficient (3.6.7) to be time-dependent.

One commonly used methodology for obtaining high-quality bottom stress measurements involves deploying closely spaced velocity sensors, beginning very near (centimeters) the bottom and extending upward a few meters (e.g., Trowbridge and Lentz, 1998). Next, for every time interval (say, every hour), a regression fit is carried out on the measured velocity profile. Then, using only the depths where points fall on a logarithmic—that is, constant-stress—profile, the fit is used to estimate $u_*$ and $z_C$ in (3.6.6).

Constant-stress layer theory clearly shows that the bottom stress is proportional to the bottom velocity squared. However, considering complex flows that include a realistic spectrum of variability presents a difficulty. Consider the simple case of a mean flow existing in an ocean with large-amplitude sinusoidal tides. If the mean stress is desired, the tides (for example) need to be included in the computation (3.6.7), because squaring the velocity means that the entire frequency spectrum contributes to amplifying the stress on a single component of the flow, for example, the mean. Wright and Thompson (1983) estimated the stress in the presence of a complicated flow field and showed that the stress due to only a component (such as the mean) of the overall flow can be linearized into the form

$$\tau^y = \rho_0 c_D \varphi v = \rho_0 \sigma_F v, \tag{3.6.10}$$

where $\varphi$ is a speed representative of the entire velocity field and is based on knowledge of the probability distribution of the overall flow. Further, $\sigma_F$ is called the *resistance coefficient*. The result is that when considering an isolated aspect of the flow (such as only its mean or low-frequency waves), it can actually be more accurate to use this linear formulation, which accounts for the complete flow field, than to use the quadratic form in (3.6.7) while neglecting the effects of the broader spectrum of variability.

Often, the classical, constant $A$, bottom boundary layer problem (as in section 3.3) is modified to include the effect of a linearized constant-stress layer. Specifically, the no-slip bottom boundary condition (3.3.1) is replaced by a stress-matching condition that joins a rotation-influenced outer Ekman layer with an inner, infinitesimally thin linearization of the logarithmic layer:

$$A(u_{Ez}, v_{Ez}) = \sigma_F(u_I + u_E, v_I + v_E) \qquad \text{at } z = -h. \tag{3.6.11}$$

where $\sigma_F$ is estimated as in (3.6.10) at a physical height of perhaps 1–2 m above the bottom. This boundary condition is often referred to as a "slip condition" because it does not require near-bottom velocity to vanish except in the limit of large $\sigma_F$. For representative parameters (e.g., $A = 5 \times 10^{-3}$ m²/s, $f = 1 \times 10^{-4}$ 1/s, $\sigma_F = 5 \times 10^{-4}$ m/s), the resulting Ekman spiral is "flattened" relative to the classic problem; that is, there is less veering.

To this point the effects of density stratification in the water column have received little attention. With strong bottom mixing, the bottom boundary layer can be treated as a mixed-layer problem analogous to that near the surface (e.g., Thompson, 1973; Weatherly and Martin, 1978). The bottom problem (at least when the bottom is flat) is simpler than the surface case in that there is no heat flux at the boundary, but it is messier in that it is often not straightforward to estimate the bottom stress given an interior velocity. Fortunately, there are simple "rules of thumb," such as Weatherly and Martin's $u_* = bc_D^{1/2}|v_I|$ (for interior flow in the $y$ direction and where $b$ is an $O(1)$ empirical constant), that appear to work tolerably well, at least in models.

## 3.7. The Bottom Boundary Layer over a Sloping Bottom

When there is an along-isobath interior flow $v_I$ in a temperature-stratified ocean, the cross-isobath flow in the bottom boundary layer leads to near-bottom warming when the Ekman flow is downslope, and cooling when the flow is upslope (Figure 3.7). This was noted in observations (e.g., Weatherly and Martin, 1978), and its implications were explored using simplified theories (MacCready and Rhines, 1991; Trowbridge and Lentz, 1991; Garrett et al., 1993). As cross-isobath advection proceeds, a horizontal density gradient (with different signs for upslope and downslope flow) forms within the bottom layer. In either case, the thermal wind equation

$$fv_z = -\frac{g}{\rho_0}\rho_x \qquad (3.7.1)$$

describes a geostrophic component of boundary layer shear (in addition to the shear associated with turbulent stresses) that acts to reduce the near-bottom along-isobath flow and hence the bottom stress. Reducing stress decreases the cross-isobath Ekman transport, thus slowing the increase in density gradient, so the deceleration process slows with time. Eventually, however, the bottom stress approaches zero, and Ekman transport also vanishes. This process of neutralizing bottom stress and Ekman transport is called *buoyancy arrest*.

As the cited authors show, the specifics of the arrest process differ considerably depending on whether the Ekman transport is upslope or downslope. These differences are explored here following Brink and Lentz (2010a), who treated the problem as a pseudo-one-dimensional process and used a turbulence closure model to parameterize the boundary layer mixing. The ambient stratification was constant, a quadratic bottom stress was used, and the interior along-isobath flow was steady and spatially uniform. They obtained numerical results and provided the following theoretical arguments, which are based largely on critical Richardson numbers, for boundary layer thickness and other properties.

With downslope Ekman transport, lighter water from upslope is pushed beneath denser ambient water, so the boundary layer thickens due to gravitational instability. Once an arrested state is reached, simple numerical experiments show that the bottom boundary layer is linearly stratified (Figure 3.8, line labeled "Downwelling") and that the corresponding along-isobath velocity $v$ has a constant vertical shear. These properties are consistent with a constant gradient Richardson number

$$Ri = \frac{-g\rho_z}{\rho_0 v_z^2}. \qquad (3.7.2)$$

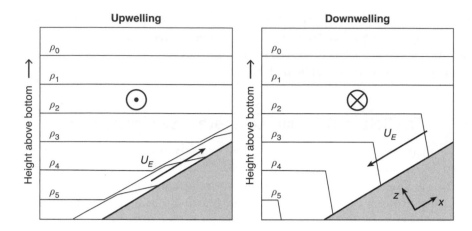

**FIGURE 3.7.** An illustration of how bottom Ekman transport affects boundary layer density. Schematics of density structure and Ekman transport for the two directions (denoted by circular symbols) of along-isobath flow. With upslope transport of dense water in the boundary layer, near-bottom stratification increases with time, while downslope transport of light water destabilizes the boundary layer, leading to anomalous thickening. Adapted from Brink and Lentz (2010b).

The total vertical density gradient in the arrested boundary layer is

$$\rho_z = \rho_{Iz} + \rho_{Ez}, \tag{3.7.3a}$$

Where $\rho_{Iz}$ is the interior (undisturbed) density gradient, and the subscript $I$ denotes a variable in the adiabatic interior. The boundary layer component $\rho_{Ez}$ is in thermal wind balance with the boundary layer component of along-isobath velocity $v_{Ez}$:

$$\rho_{Ex} = -f\rho_0 v_{Ez}/g. \tag{3.7.3b}$$

For total velocity $v_I + v_E$ to be continuous at the top of the boundary layer,

$$v_E = v_I (z - \delta_d)/\delta_d, \tag{3.7.3c}$$

where $\delta_d$ is the arrested, restratified boundary layer thickness (the subscript $d$ is a reminder that this is for the case with a downslope flow before arrest occurs). Assuming that both the interior velocity and vertical density gradient are spatially uniform, a coordinate rotation shows that

$$\rho_{Ex} = -\alpha\rho_{Ez}, \tag{3.7.4a}$$

where $\alpha$ is the constant bottom slope. Combining (3.7.3b, 3.7.3c, and 3.7.4a) leads to the total near-bottom vertical density gradient of

$$\rho_z = \rho_{Iz} + f v_I \rho_0 / (\alpha g \delta_d). \tag{3.7.4b}$$

This form, along with (3.7.3c) is then used in the definition of the critical Richardson number (3.7.2) to obtain a single equation for $\delta_d$,

$$0 = \delta_d^2 N^2 \alpha - \delta_d v_I f - v_I^2 \alpha \, Ri, \tag{3.7.5}$$

which yields

$$\delta_d = \frac{v_I}{sN}\Gamma,$$ (3.7.6a)

where

$$\Gamma = \frac{1}{2}[1 + (1 + 4Ri_d s^2)^{1/2}] \geq 1,$$ (3.7.6b)

and empirically from model runs, $Ri_d = 0.7$ ($d$ denotes the value for downwelling). The *slope Burger number* (which can be thought of as a scaled bottom slope) is

$$s = \frac{\alpha N}{f}$$ (3.7.6c)

The adjustment time scale to reach this state is found (based on a scaling argument and using a least-squares fit to obtain an empirical constant) to be

$$T_d = \Gamma(1 + s^2)(1.6\,fb^2 d\,s^3)^{-1},$$ (3.7.7a)

where the frictional parameter is

$$d = \frac{c_D N}{f}.$$ (3.7.7b)

The empirical constant $b$ ($= 0.63$) in (3.7.7a) relates the interior velocity to the bottom stress when the bottom is flat, as in Weatherly and Martin (1978):

$$\tau_B^y = \rho_0 b^2 c_D |v_I| v_I.$$ (3.7.8)

More refined, three-dimensional calculations that resolve eddies within the bottom boundary layer provide much greater insight with regard to the actual turbulent processes but otherwise agree with the Brink and Lentz findings (e.g., Wenegrat and Thomas, 2020). Perhaps this is not surprising in this case, since a turbulence closure scheme is presumably likely to work well in the downwelling case where the turbulence is driven by straightforward gravitational instability, as opposed to potentially more complex shear instabilities.

A similar approach applied to an upwelling boundary layer leads to more complicated results. For smaller $s$, the arrested bottom boundary layer has a sharply defined cap ("Capped" in Figure 3.8), below which the boundary layer is stably stratified. In this case, the boundary layer thickness was given to a good approximation by Middleton and Ramsden (1996):

$$\delta_{uc} = \frac{1.4u_*}{\sqrt{fN}}(1 + s)^{-1}.$$ (3.7.9)

Notice the similarity of this form to that for a surface mixed layer (3.5.10) which is also bounded by a sharp density contrast. Indeed, the sloping bottom only leads to a

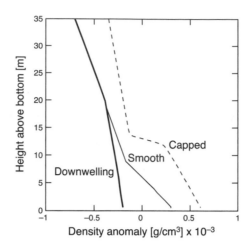

**FIGURE 3.8.** Representative near-bottom density profiles for three cases with complete buoyancy arrest, computed using a pseudo-one-dimensional numerical model with a turbulence closure scheme. Downwelling ($s = 1.38$), capped upwelling ($s = 0.33$), smooth upwelling ($s = 1.38$). Note that to achieve a small value of $s$, the vertical density gradient in the interior is weaker in the capped upwelling case. Adapted from Brink and Lentz (2010a).

correction to the flat-bottom result in the capped parameter range. The adjustment time scale to arrest is

$$T_{uc} = (1 + s^2)[(1 + s)\ s^2 f d^{1/2}]^{-1}. \tag{3.7.10}$$

The capped results are consistent with the notion that in the limit of a flat bottom ($s = 0$), there should be a simple bottom mixed layer and no buoyancy arrest.

For larger values of $s$, there is no density jump across the top of the upwelling bottom boundary layer, but the waters within the arrested layer are strongly stratified ("Smooth" in Figure 3.8). This is called the "smooth" case, and an expression for its thickness is found using an argument parallel to that used in the downwelling case (leading to 3.7.6), namely,

$$\delta_{us} = \left( \frac{-v_I}{sN} \right) \Lambda, \tag{3.7.11a}$$

where

$$\Lambda = \frac{1}{2}[-1 + (1 + 4Ri_u s^2)^{1/2}] \geq 0, \tag{3.7.11b}$$

with $Ri_u = 0.4$. Finally, the adjustment time scale in this smooth case is

$$T_{us} = \frac{1.2(1 + s^2)\Lambda}{b^2 s^3 df}. \tag{3.7.12}$$

In practice, the greater of the two depths (smooth or capped) is the correct choice. Comparing (3.7.11b) with (3.7.6b), it is clear that the downwelling boundary layer is thicker (all else being the same) than at least the smooth upwelling case, and the capped case is

**TABLE 3.1.** Comparison of Representative Summertime Arrested Bottom Boundary Layer Properties

|  | Oregon Shelf | Oregon Slope | Middle Atlantic Bight Shelf | Peru Shelf |
|---|---|---|---|---|
| $\alpha$ | 0.007 | 0.034 | 0.001 | 0.01 |
| $s$ | 0.5 | 1 | 0.1 | 1 |
| $d$ | 0.3 | 0.3 | 0.3 | 0.6 |
| $\delta_d$ (m) | 23 | 15 | 101 | 15 |
| $\delta_u$ (m) | 4 capped | 3 smooth | 4 capped | 4 capped |
| $T_d$ (days) | 7 | 2 | 610 | 2 |
| $T_u$ (days) | 0.5 | 0.7 | 13 | 0.3 |
| $L_A$ (km) | 2100 | 800 | 42,000 | 650 |

generally thinner than the downwelling result as well, unless $d$ is substantially greater than unity. This thickness disparity (depending on the sign of $v_I$) is consistent with the notion that downwelling leads to gravitational instability, while upwelling strengthens near-bottom stratification and so inhibits mixing.

These various expressions illustrate a few important points. One is that the arrested boundary layer thickness $\delta$ always depends on the slope Burger number $s$, although the dependence may become fairly weak, as in the downwelling case with larger $s$. The adjustment time scales, however, all depend on $d$, and very strongly on $s$. It is useful to compare a few representative summertime continental margin settings (Table 3.1): Oregon (both shelf and slope), Middle Atlantic Bight (the shelf south of New England), and Peru near 15°S, all computed using $|v_I| = 10$ cm/s. Over the very gently sloping ($\alpha \approx 0.001, s = 0.1$) Middle Atlantic Bight shelf, adjustment times are much longer than in the other cases, and it seems unlikely that downwelling arrest could ever occur given realistically variable currents and stratification. The other geographical settings (both upwelling and downwelling) have plausible thicknesses and have time scales comparable to, or shorter than, the time scales of synoptic wind driving (which is often a major driver for shelf current fluctuations).

To this point it has been assumed that everything about the problem is uniform along isobaths. This is inconsistent with observed oceanic conditions, where at least some alongshore variation is ubiquitous. Chapman and Lentz (1997) and Brink (2012a) looked into how a steady flow entering the shelf unequilibrated to buoyancy arrest adjusts over distance alongshore.[4] For the case of a depth-independent interior flow, the alongshore scale of this adjustment is given by

---

4 The problem could be thought of as being like the "arrested topographic wave" (section 5.11) but including stratification and allowing buoyancy arrest to develop.

$$L_A = 39 \frac{(1+s^2)}{ds^2} \left( \frac{|Q|}{fh} \right)^{1/2} \tag{3.7.13}$$

(Brink, 2012a) regardless of the alongshore flow direction. In this case, $Q$ is the alongshore volume transport (e.g., m³/s) and $h$ is a representative water depth. If a flow configuration has an along-isobath scale (due to eddies, topographic irregularity, or forcing, for example) much larger than $L_A$, arrest can go to completion, but configurations having scales comparable to, or shorter than $L_A$ cannot achieve complete buoyancy arrest even after extended times. The representative values for $L_A$ in Table 3.1 (computed using $Q = 1 \times 10^5$ m³/s and $h = 100$ m) are all considerably larger, for example, than midocean mesoscale eddies (around 100 km) which might impinge at the shelf edge. Given known topographic irregularities and alongshore variations in wind forcing, it seems that there are probably few contexts in the coastal ocean which are sufficiently uniform to allow complete arrest ever to occur.

What do actual ocean observations reveal about buoyancy arrest? Early efforts (e.g., Weatherly and Martin, 1978) established that upslope Ekman transport correlates with near-bottom cooling and vice versa. Further, Lentz and Trowbridge (1991) demonstrated a clear tendency for thicker bottom boundary layers with poleward flow (downslope Ekman transport) and vice versa on the shelf north of San Francisco (Figure 3.9). Both findings are consistent with the physics of buoyancy arrest but do not demonstrate that complete arrest occurs. Lentz and Trowbridge (2001) used long time series of currents and stress from a northern California midshelf site where the mean flow is poleward. On time scales of a week or longer, they found a clear tendency for along-isobath flow to be in thermal wind balance as close to the bottom as 1 m, again consistent with the physics of buoyancy arrest. They estimated a time scale for downwelling buoyancy arrest to be 6 days using an expression similar to (3.7.7a), a result roughly consistent with the observations. It appears that the bottom stress is substantially reduced owing to the thermal wind, but because of ambiguities about the definition of bottom Ekman transport, the authors stopped short of saying that complete buoyancy arrest has occurred. Certainly, it is safe to say that the underlying physics of Ekman-induced thermal wind shear leading to reduced bottom stress is at play even if it does not go to completion.

## 3.8.   Conclusion

Surface and bottom boundary layer processes are important anywhere in the ocean: they are the contact zones between the ocean and its surroundings. The boundary layers are particularly important over the continental shelf where they can occupy much, or even all, of the water column. Indeed, over the inner shelf, the entire water column is a turbulent boundary layer, a situation special enough that it requires a good deal more consideration (chapter 6). Thus, variability in boundary layer thickness, which may not be important in 4000 m of water, becomes an important concern when the water is 100 m deep. Further, the boundary layers can be difficult places to make measurements, both

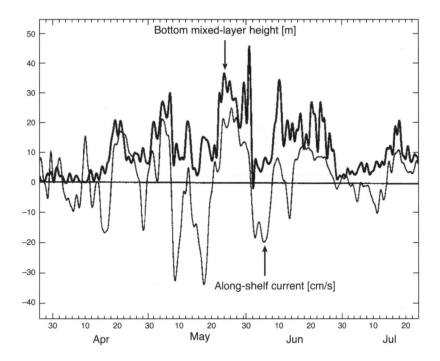

**FIGURE 3.9.** The relation of bottom boundary layer thickness to the direction of alongshore flow, hence Ekman transport. Alongshore velocity (lighter line) and bottom mixed-layer thickness (heavier line) at midshelf north of San Francisco. Records have been smoothed to remove tides and other higher-frequency processes. Note that for a given magnitude of current, the bottom mixed layer is thicker for positive (poleward) flow and thus downslope Ekman transport than for negative velocity; for example, compare thickness on May 26 with that on May 18. Adapted from Lentz and Trowbridge (1991).

in the upper ocean (where mooring motion and breaking waves are issues) and near the bottom (where tight spacing is required to resolve stress profiles, and where sediment transport events can change the bottom).

Boundary layer processes remain a very active subject for investigation. A good deal of upper-ocean research is ongoing, especially in an open-ocean context, where the effects of lateral gradients and submesoscale instabilities are of great interest (e.g., Dong, et al., 2020). Also, there is a growing appreciation of Langmuir turbulence (e.g., Thorpe, 2004), where the Stokes drift associated with surface waves (section 6.3) is central to generating a new class of overturning circulation. This process appears to play a substantial role in upper-ocean mixing on scales up to the global (Belcher, et al., 2012). At depth, there is growing evidence that the peculiar properties of Ekman layers over a sloping bottom, e.g., a midocean ridge, may play a role in governing basin-scale circulation (e.g., Drake et al., 2020). Further, there is increasing evidence that baroclinic instabilities also play a role near the bottom (section 9.4). Thus, we can anticipate an active stream of new and relevant results for years to come.

# 4

# Two-Dimensional Models
# of Wind Forcing

## 4.1. Introduction: Coastal Upwelling

Smith (1968) provides an interesting historical context for the study of wind driving in the coastal ocean. It seems that even in early colonial times, people recognized that waters off the west coast of the Americas are quite cool in the summertime. Indeed, this is obvious to even casual tourists in San Francisco to this day. For years, it was believed that there were simply cold alongshore currents arriving from high latitudes—certainly a reasonable supposition. However, by the early- to mid-1800s enough information had been gathered for it to be apparent that counter to this assumption, waters actually become warmer poleward off the west coast (e.g., Figure 4.1), ruling out a continuous cold current as an explanation. It then became clear that the only good resolution is that colder, deeper water is rising toward the ocean surface near the coast.

Although winds were suspected as a driving agent for this vertical motion, the mechanism was not clear until Ekman (1905) developed his boundary layer model (section 3.2) and Thorade (1909) applied those ideas. Simply stated, summertime equatorward alongshore winds, in conjunction with Earth's rotation, give rise to an offshore transport of surface waters, and these are replaced by colder waters from below. There is now a vast observational literature consistent with this finding. Particularly important regions of summertime coastal upwelling (e.g., Mackas et al., 2005) include the midlatitude west coasts of South and North America, off southwest and northwest Africa, Portugal, and the Arabian coast of the Indian Ocean. (This list is far from complete and ignores many settings, including where the wind forcing does not follow the same seasonal pattern.) However, the Ekman/Thorade dynamical picture is incomplete in that it does not deal with a number of important effects, including bottom topography and density stratification.

Before proceeding with models, it helps to review some observations. Typically, during the summertime upwelling season off Oregon, there is a band of cold surface water

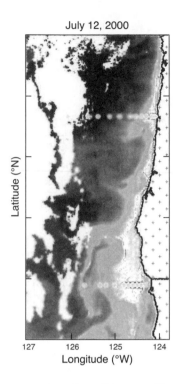

July 12, 2000

**FIGURE 4.1.** Satellite sea surface temperature image from off the coast of Oregon and northern California. July 12, 2000, where the warmest water is the darkest shade, and the coolest water is the lightest. Note that the water is generally warmer offshore and toward the north. White areas represent clouds. The latitude range is 41°–46° N. Adapted from Huyer et al. (2005).

that can extend for more than 100 km offshore (Figure 4.1). Often, there is a distinct surface front, with a temperature jump of a few degrees over a distance of less than about 10 km (Figure 4.2). Looked at in sectional form, it is clear that the cold, salty nearshore surface water is continuous with water offshore of the front at depths of about 100 m. Generally in the ocean, near-surface waters (shallow enough for light to penetrate) are depleted in nutrients, while deeper waters contain nutrients but not the light that microscopic plants (phytoplankton) need to thrive. Upwelling regions are a special environment, because the vertical velocity carries nutrients upward to where the light is available. The net result is lush phytoplankton growth and a productive overall ecosystem. Figure 4.2 (third panel) shows the chlorophyll a pigment concentration, a useful proxy for phytoplankton biomass. In addition, the upwelled waters cool the lower atmosphere and thus affect cloudiness and winds along the coast (e.g., Beardsley et al., 1987). These observations raise numerous questions, for example, about the properties of the upwelling front, the distribution of vertical velocity, and the fate of upwelled water.

This chapter builds on an understanding of boundary layer processes and considers exclusively models that are two-dimensional in the offshore-vertical $(x, z)$ plane, which requires that bottom topography vary only offshore (and not alongshore) and that

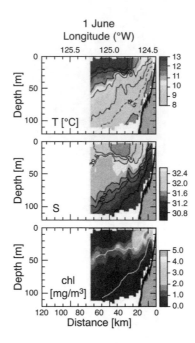

**FIGURE 4.2.** Cross-shelf sections of temperature (top), salinity and chlorophyll a concentration (bottom) during coastal upwelling off Oregon at 43.22° N, June 1, 2000. Note that at the surface, the coolest, saltiest water occurs nearshore, and the highest chlorophyll concentration is found at the front (where the horizontal temperature gradient is sharp). Adapted from Barth et al. (2005).

currents have infinite alongshore scales. What does "infinite" mean in this context? Physically, it means that alongshore variations must occur over a scale large enough that frictional effects can erase any knowledge of these distant variations. An alongshore frictional scale $L^F$ can be found by scaling a linearized version of (2.4.10b). Specifically, the alongshore pressure gradient is compared with the bottom stress. Further, the bottom stress is estimated by the simple form (3.6.10), and the alongshore velocity is assumed to be in geostrophic balance. Alongshore variations can then be ignored if

$$L^y \gg \frac{fhL^x}{\sigma_F} = L^F, \tag{4.1.1}$$

where $L^y$ is the actual alongshore scale of topography, winds, or currents; $f$ is the Coriolis parameter; $h$ is the local water depth; $L^x$ is a cross-shelf scale (that could be taken to be the distance from the coast); and $\sigma_F$ is a bottom resistance parameter. For a representative midshelf situation ($L^x = 30$ km, $h = 75$ m, $\sigma_F = 5 \times 10^{-4}$ m/s, and $f = 1 \times 10^{-4}$ 1/s), ignoring alongshore variations requires that $L^y \gg 450$ km. For comparison, shelf-slope topography is never smooth over hundreds of kilometers, and weather systems that drive currents typically have scales of $O(1000$ km). Thus, two-dimensional models are most applicable where the water is shallower.

There is sometimes confusion about the physical meaning of a two-dimensional model. Because alongshore variability (or any information about distant happenings) does not enter such a model explicitly, these models are sometimes treated as describing entirely local processes. This is directly contrary to the reality that these models represent the response to forcing on the very largest alongshore scales. All told, the two-dimensional assumption is probably never completely valid at midshelf or farther offshore, but it is an extremely valuable starting point for understanding a three-dimensional world. These simplified models elucidate the fundamental underlying processes of wind driving and thus help rationalize some important observations.

## 4.2.   Governing Equations for a Barotropic Ocean

The simplest way to start is by ignoring density stratification and considering a barotropic (constant-density) coastal ocean. The governing equations when flow is uniform along-shore (i.e., $u_y$, $v_y = 0$) are

$$u_t + u u_x + w u_z - f v = -\frac{1}{\rho_0} p_x + (A u_z)_z \tag{4.2.1a}$$

$$v_t + u v_x + w v_z + f u = -\frac{1}{\rho_0} p_y + (A v_z)_z \tag{4.2.1b}$$

$$u_x + w_z = 0 \tag{4.2.1c}$$

Horizontal viscosity has been ignored, the $x$ direction and corresponding velocity component $u$ are positive offshore, $(y, v)$ are alongshore, and $(z, w)$ are vertical. Subscripts with regard to independent variables ($x, y, z$, and time $t$) represent partial differentiation. The constant Coriolis parameter is $f$, $\rho_0$ is a constant density, $h(x)$ is the depth, pressure is $p$, and $A$ is a vertical eddy viscosity. With the assumption of a small Rossby number $\hat{u}/(fL)$ (where $\hat{u}$ is a representative velocity magnitude, and $L$ is a representative cross-shelf scale), the nonlinear terms are neglected for the time being:

$$u_t - f v = -\frac{1}{\rho_0} p_x + (A u_z)_z, \tag{4.2.2a}$$

$$v_t + f u = -\frac{1}{\rho_0} p_y + (A v_z)_z, \tag{4.2.2b}$$

$$u_x + w_z = 0. \tag{4.2.2c}$$

At this point, as in section 3.1, it is convenient to divide the velocity components into "interior" (subscript $I$) and boundary layer (subscript $E$) components, so that, for example, $u = u_I + u_E$, where the interior equations are

$$u_{It} - f v_I = -\frac{1}{\rho_0} p_x, \tag{4.2.3a}$$

$$v_{It} + fu_I = -\frac{1}{\rho_0} p_y, \tag{4.2.3b}$$

$$u_{Ix} + w_{Iz} = 0, \tag{4.2.3c}$$

and the boundary layer equations are

$$u_{Et} - fv_E = (Au_{Ez})_z, \tag{4.2.4a}$$

$$v_{Et} + fu_E = (Av_{Ez})_z, \tag{4.2.4b}$$

$$u_{Ex} + w_{Ez} = 0. \tag{4.2.4c}$$

Note that the boundary layer equations do not explicitly depend on $x$, although horizontal dependence can occur through the surface and bottom boundary conditions. Also, the interior horizontal velocity components are independent of $z$.

The alongshore pressure gradient $p_y$ in (4.2.3b) deserves comment. It is straightforward to use (4.2.3) to obtain an expression for $u_I$, hence $u_{Iy}$, which must then vanish because currents are taken to be uniform alongshore. This, in turn, requires that $p_{yy}$ and $p_{xy}$ both vanish. The latter statement requires that the alongshore pressure gradient, which must be imposed in the present context (rather than found as part of the solution), cannot vary across the shelf. Stated another way, any geostrophic cross-shelf velocity $u_G$ associated with the alongshore pressure gradient has to be constant across the entire domain. This is a rather strong, restrictive constraint in view of the transport $hu_G$, which could become huge in deep water for even a small $u_G$. In contrast, for a truly three-dimensional problem where $u_{Iy} \neq 0$, there is no such constraint on $p_y$; that is, $p_{xy}$ can be nonzero (e.g., section 5.7). From this point on, the alongshore pressure gradient will be set to zero, but the fact remains that a two-dimensional $(x, z)$ mass balance does not require pressure to be perfectly uniform alongshore.

Now, return to the general statement for total velocity (4.2.2). The continuity equation can be depth-integrated using the linearized ($\zeta/h \ll 1$) surface boundary condition

$$w = \zeta_t \qquad \text{at } z = 0 \tag{4.2.5a}$$

and no flow through the bottom:

$$w = -h_x u \qquad \text{at } z = -h(x). \tag{4.2.5b}$$

The free-surface elevation $\zeta = p/(g\rho_0)$ because of the hydrostatic balance. The depth-integrated continuity equation now becomes

$$U_x + \zeta_t = 0, \tag{4.2.6a}$$

where the transport

$$U = \int_{-h}^{\zeta} (u_I + u_E) dz \cong hu_I + \int_{-h}^{0} u_E dz = U_I + U_E \tag{4.2.6b}$$

for $h \gg |\zeta|$.

Returning to the boundary layer problem, it is now useful to assume a constant eddy viscosity $A$ and that time variations occur on scale $\hat{t}$ (a few days or longer) that is long relative to the inertial period; that is, $|f\hat{t}| \gg 1$. The time scale assumption greatly simplifies the boundary layer problem (4.2.4) by allowing the time derivatives to be dropped. This assumption is not required; the system can still readily be solved with sinusoidal time dependence (e.g., Book et al., 2009), but the algebra is a good deal messier. In any case, the boundary conditions are

$$\rho_0 A(u_z, v_z) = (\tau_0^x, \tau_0^y) \qquad \text{at } z = 0, \qquad\qquad (4.2.7a)$$

and (consistent with a constant $A$)

$$u_I + u_E = 0, \qquad\qquad\qquad\qquad\qquad\qquad (4.2.7b)$$

$$v_I + v_E = 0, \qquad \text{at } z = -h(x), \qquad\qquad (4.2.7c)$$

where $(\tau_0^x, \tau_0^y)$ is the surface wind-stress vector. Consistent with the observed dominance of alongshore winds off the U.S. west coast, the cross-shelf wind stress $\tau_0^x$ is ignored for now, but it is revisited in sections 5.7 and 6.2. It is now straightforward to solve the slowly varying boundary layer problem. It is also assumed that the Ekman scale depth (sections 3.2 and 3.3)

$$\delta_E = \left( \frac{2A}{|f|} \right)^{1/2} \qquad\qquad\qquad\qquad\qquad (4.2.8)$$

is substantially less than the water depth everywhere. There is, however, an artificial coastal wall, perhaps 30 m high at $x=0$, so that $3\delta_E/h \leq 1$ everywhere (assuming that depth increases offshore), an idealization justified by Mitchum and Clarke (1986a) for situations where the alongshore scale is much greater than the cross-shelf scale. A more realistic approach to the problem, which avoids inserting this barrier, is addressed in section 6.2. Constant-$A$ boundary layer problems such as this are sometimes solved with a bottom condition (3.6.11) that more closely resembles coupling to a logarithmic layer (section 3.6) rather than the no-slip condition (4.2.7b, c) employed here. Using a constant eddy viscosity and no-slip boundary conditions may not be particularly realistic, but the results are meant to be illustrative.

Now, consider the interior equations (4.2.3) again. With $p_y = 0$, scaling the remaining terms in (4.2.3b) shows that $|v_I|/|u_I| = O(f\hat{t})$, that is to say that interior alongshore flow is much stronger than cross-shelf velocity. This, in turn, means that the $u_{It}$ term in (4.2.3a) is doubly small, $O[(f\hat{t})^{-2}]$ compared with the $fv_I$ term, so it can be safely dropped. Given these assumptions, the interior momentum equations are simply

$$-fv_I = -\frac{1}{\rho_0} p_x, \qquad\qquad\qquad\qquad\qquad (4.2.9a)$$

$$v_{It} + fu_I = 0. \qquad\qquad\qquad\qquad\qquad\qquad (4.2.9b)$$

An important simplification can now be obtained by scaling the continuity equation (4.2.6a) using $U_I = hu_I$ and (4.2.9). The surface elevation term in (4.2.6a) can then be neglected when

$$D = (fL^x)^2/gH \ll 1. \qquad\qquad\qquad\qquad\qquad (4.2.10)$$

Physically, this condition means that the barotropic *Rossby radius of deformation*, $\sqrt{gH}\,/\,f$, is large relative to a typical cross-shelf scale $L^x$, which might be set by the bottom topography or wind variations, for example. The Rossby radius is a typical scale of surface deflection for divergent flow in a flat-bottom ocean and might be around 500 km for $H = 100$ m at midlatitudes, in comparison with a representative shelf width of perhaps 75 km. This result, that flow divergence is negligible, is called the *rigid-lid approximation*.[1] An important outcome of this follows because there can be no net flow through a coastal boundary at $x = 0$, so that $U = 0$ there, and when $U_x = 0$ (neglecting divergence in 4.2.6a), then $U = 0$ everywhere. Note that this does not mean that $u_I$ or $u_E$ vanishes everywhere but only that $U_I = -U_E$, so the depth integral (4.2.6b) vanishes everywhere.

A few words about the rigid-lid approximation are in order. When it applies, it means that the volume storage associated with the up-and-down motion of the free surface is negligible. The free surface still moves, however, and thus still governs the near-surface pressure field. Stated another way, there is nothing about making the approximation that inhibits the actual motion of the free surface.

To proceed, (4.2.9b) can be vertically integrated and the condition that $U = 0$ applied, so that

$$hv_{It} = -fU_I = fU_E. \tag{4.2.11a}$$

Assuming a constant eddy viscosity and that the boundary layers are thin relative to the water depth, $\delta_E/h \ll 1$, we can apply expressions (3.2.11) and (3.3.4) for the surface and bottom Ekman transports, respectively, to obtain

$$U_E = (\rho_0 f)^{-1}\tau_0^y - v_I\delta_E/2 - u_I\delta_E/2 \cong (\rho_0 f)^{-1}\tau_0^y - v_I\delta_E/2. \tag{4.2.11b}$$

The $u_I$ term is dropped because of the assumption of slow time variability.

## 4.3. Midshelf Response to Wind Forcing

It is useful to look more closely now at the response of a constant-density ocean to alongshore winds over the midshelf. The governing equation, based on (4.2.11),

$$hv_{It} = \rho_0^{-1}\tau_0^y - v_I\delta_E f/2, \tag{4.3.1}$$

can be thought of as expressing a balance of cross-shelf surface Ekman transport

$$U_{ES} = (f\rho_0)^{-1}\tau_0^y, \tag{4.3.2a}$$

bottom Ekman transport

$$U_{EB} = -\frac{v_I\delta_E}{2}, \tag{4.3.2b}$$

---

1 A more general way to derive this approximation is to ignore dissipation and then derive a depth-averaged vorticity equation; that is, form a single equation for pressure from (4.2.3) and using (4.2.5). Scaling that equation then leads to this condition in a straightforward way that does not require slow time variations (see section 5.5).

and (from 4.2.11a) flow balancing the interior alongshore acceleration through the Coriolis force

$$U_I = -hv_{It} / f. \tag{4.3.2c}$$

Thus, (4.3.1) can be rewritten as a statement that there is no net cross-shelf transport:

$$0 = U_I + U_{ES} + U_{EB}. \tag{4.3.3}$$

As an aside, notice that an across-shelf wind stress in deep water gives rise (through 3.2.11) to only an alongshore Ekman transport, which would not enter (4.3.3); this alongshore boundary layer transport goes on uninterrupted indefinitely if there is no cross-shelf barrier. In contrast, alongshore winds drive a cross-shelf Ekman transport $U_{ES}$ that encounters the coastal wall and thus needs to be balanced so that there is no net cross-shelf volume flux at the boundary. When $|f\hat{t}| \gg 1$, cross-shelf wind stress is an effective driving agency only on the inner shelf (section 6.2) or where the alongshore scale of the cross-shelf wind is short enough to generate a substantial wind-stress curl, as is found with hurricanes or topographically channeled wind jets (e.g., McCreary et al., 1989; Trasviña et al., 1995). This latter possibility, of course, violates the two-dimensional assumption.

Consider, now, how the ocean responds to the abrupt onset of a steady alongshore wind stress,

$$\tau_0^y = 0 \qquad \text{for } t < 0, \tag{4.3.4a}$$

$$\tau_0^y = \tau_A \qquad \text{for } t > 0, \tag{4.3.4b}$$

with $v_I = 0$ at $t = 0$. The solution to (4.3.1) is then

$$v_I = \frac{2\tau_A}{\rho_0 \delta_E f} \left[ 1 - \exp\left( -\frac{t}{t_f} \right) \right] \tag{4.3.5a}$$

for $t \geq 0$, where the frictional spindown time is

$$t_f = \frac{2h(x)}{\delta_E f} = \frac{h(x)}{\sigma_F}. \tag{4.3.5b}$$

Pressure can be found by integrating the geostrophic relation (4.2.9a) in $x$. The solution (4.3.5) does not include any near-inertial oscillations (see section 5.12) because $u_{It}$ is ignored in (4.2.9a), consistent with using the long time-scale assumption, $|f\hat{t}| \gg 1$. For the same reason, the Ekman layer appears to be established instantaneously. The associated cross-shelf transports (4.3.2) are then

$$U_{ES} = (f\rho_0)^{-1} \tau_A, \tag{4.3.6a}$$

$$U_{EB} = -\frac{\tau_A}{\rho_0 f} \left[ 1 - \exp\left( -\frac{t}{t_f} \right) \right], \tag{4.3.6b}$$

$$U_I = -\frac{\tau_A}{\rho_0 f} \exp\left( -\frac{t}{t_f} \right). \tag{4.3.6c}$$

This result clarifies how the no-transport requirement (4.3.3) is met. The alongshore wind drives a surface Ekman transport $U_{ES}$ which is constant over all time $t > 0$. Initially, this is balanced by an interior transport $U_I$, which, in turn, is tied, through the Coriolis force, to the acceleration of an alongshore flow. As the alongshore flow strengthens, it generates a growing bottom Ekman transport $U_{EB}$. The bottom Ekman transport gradually eliminates the need for an interior transport. Stated another way, the interior alongshore flow accelerates over a time scale $t_f$ until it reaches a constant value, and $U_I = 0$. The net effect is that the wind generates an alongshore flow, but this does not happen directly, because the stress does not act directly on the interior water column. Rather, acceleration occurs through the agency of the interior cross-shelf flow (via 4.3.2c), which itself is initially required because of the coastal boundary condition. This cross-shelf flow in reality passes though the inner shelf, where it is deflected vertically to balance the surface Ekman transport (section 6.2). In the final steady state the surface stress is balanced by the bottom stress everywhere, consistent with $U_{ES} + U_{EB} = (\tau_0^y - \tau_B^y)/(f\rho_0) = 0$.

The frictional adjustment time (4.3.5b) is not constant but is shorter in shallower water. This is consistent with notions of the spinup time depending on the volume (or depth) of water that needs to accelerate. Physically, a steady state is established earlier in shallow water (perhaps over a time scale of about 5 days in 100 m of water) than in deep water, where the time scale can be hundreds of days in an ocean basin. If the forcing were sinusoidal in time, the analogous result would be that winds and alongshore currents would be much more nearly in phase in shallow water than in deeper waters. Geostrophy (4.2.9a) demands that the pressure gradient $p_x$ at a given location adjusts on the same time scale as the local $v_I$, but pressure itself responds on a time scale as long as that for $v_I$ in waters farthest offshore.

One further aspect of this solution gives pause: for long time scales, the cross-shelf flow that balances the surface Ekman transport occurs entirely in the bottom boundary layer everywhere. This means that for upwelling-favorable winds, the upwelled water eventually comes from near the bottom, no matter how deep the water might become offshore. Hydrographic observations (e.g., Huyer et al., 1987, 2005), however, suggest that water upwelled at the coast typically originates at depths of 150 m or less. It is tempting to ascribe this stark contrast simply to the inhibition of vertical motions by density stratification, but this idea is ruled out in the following sections.

## 4.4.    Inclusion of Moderate Density Stratification

Now, consider an ocean with horizontally uniform stratification, and again, assume that the alongshore scale for velocity and density is infinite; that is, $u_y = v_y = \rho_y = 0$. The linear equations of motion for the inviscid interior (outside the turbulent boundary layers) are thus

$$u_t - fv = -\frac{1}{\rho_0} p_x, \qquad\qquad (4.4.1a)$$

$$v_t + fu = -\frac{1}{\rho_0} p_y, \tag{4.4.1b}$$

$$0 = -p_z - g\rho_2, \tag{4.4.1c}$$

$$u_x + w_z = 0, \tag{4.4.1d}$$

$$0 = \rho_{2t} + w\rho_{1z}, \tag{4.4.1e}$$

where density

$$\rho = \rho_0 + \rho_1(z) + \rho_2(x,z,t), \tag{4.4.1f}$$

and

$$\rho_0 \gg |\rho_1(z)| \gg |\rho_2(x,z,t)|. \tag{4.4.1g}$$

Again, to maintain alongshore uniformity, $p_{xy} = p_{yy} = p_{zy} = 0$, where the last equality stems from (4.4.1c). In addition, attention is again restricted to time scales $\hat{t}$ long relative to the inertial, $|f\,\hat{t}\,| \gg 1$, so that (4.4.1a) reduces simply to geostrophy. With this simplification, it is then straightforward to eliminate $v$ from (4.4.1a,b) and $\rho_2$ from (4.4.1c, e) to obtain from (4.4.1d)

$$0 = p_{xxt} + f^2 \left( \frac{p_{zt}}{N^2} \right)_z, \tag{4.4.2}$$

where $N^2$ is the buoyancy frequency squared, $-\dfrac{g\rho_{1z}}{\rho_0}$. Forcing and dissipation enter the problem through boundary conditions expressing, for example, Ekman pumping. The first term in (4.4.2) is proportional to the time change in relative vorticity, $v_{xt}$. The second term represents the time change in vertical distance between isopycnals, that is, stretching within the water column. Thus, the physical meaning of this equation is that changes in relative vorticity are balanced by vortex stretching.

To proceed, boundary conditions are required. For simplicity, the surface and bottom Ekman layers are assumed to be infinitesimally thin, and only an alongshore wind stress is applied. At the surface (using 3.4.4 and 4.2.5a),

$$w = \frac{1}{f\rho_0}(\tau_0^y)_x + \frac{1}{g\rho_0} p_t \qquad \text{at } z = 0, \tag{4.4.3a}$$

which reduces to $w = 0$ for spatially uniform winds and a rigid lid. At the bottom (see section 3.4),

$$w = -h_x u + \frac{1}{f\rho_0}(\tau_B^y)_x \qquad \text{at } z = -h(x), \tag{4.4.3b}$$

where $\tau_B^y$ is the alongshore component of the bottom stress, which is, in turn, dependent on the bottom velocity. At the coastal wall, extremely thin boundary layers are assumed, so that

$$0 = u + U_{ES}\delta_D(z + \varepsilon) + U_{EB}\delta_D(h - \varepsilon + z) \qquad \text{at } x = 0, \qquad (4.4.3\text{c})$$

where $\delta_D$ is the Dirac delta function, ($\delta_D(\xi) = 0$ everywhere except at $\xi = 0$, and the integral of $\delta_D$ over all $\xi$ is 1). The very small depth increment $\varepsilon$ is chosen to remove any ambiguities about whether $\delta_D$ is to be treated as inside or outside the water column. Far from shore, the solution is not allowed to blow up (i.e., it is bounded).

Scaling the vorticity equation (4.4.2) is illuminating. Say that the horizontal length scale is $L^x$, the vertical length scale is $L^z$, and that $N_0^2$ is a typical squared buoyancy frequency. It then follows that the vertical scale (often called the deformation scale) is given by

$$L^z = \frac{f}{N_0} L^x. \qquad (4.4.5)$$

Thus, flow with a large cross-shelf scale has a large vertical scale, so that velocity is relatively depth-independent. Since the coastal wall at $x = 0$ is usually taken to be in shallow water, it is often reasonable to assume that $L^z$ is much greater than the water depth at the wall, so that $u_{Iz} \approx 0$ at $x = 0$, which allows condition (4.4.3c) to be simplified by depth-integrating.

Nondimensionalizing (4.4.2) leads to

$$0 = Bu \; p'_{x'x't'} + \left( \frac{p'_{z't'}}{N'^2} \right)_{z'} \qquad (4.4.6)$$

where primed entities here are nondimensional, and the *Burger number*

$$Bu = \frac{N_0^2 H^2}{f^2 L^{x2}} \qquad (4.4.7)$$

can be thought of as the squared ratio of the internal Rossby radius of deformation (of order $N_0 H/f$) to the width of the shelf-slope topography. (For this purpose, it is reasonable to take $H$ to be the maximum water depth.) The Burger number can also be interpreted as the squared ratio of the water depth to the natural vertical scale $L^z$ (4.4.5) over which pressure varies (i.e., the deformation scale).

When the stratification is weak in the sense of small $Bu$, a perturbation solution to (4.4.6) is possible (Clarke, 1976; Janowitz and Pietrafesa, 1980). With steady wind forcing applied suddenly (4.3.4), the lowest order, $O(Bu^0)$, interior solution is depth-independent and identical to (4.3.5a), and no density variations occur. The first, $O(Bu)$, correction is depth-dependent and is driven by the lowest-order vertical velocity. Because $w' = 0$ at the surface, lowest-order $w$ is strongest at the bottom (4.4.3b), and so the $O(Bu)$ density variations are strongest at the bottom when $N'$ is constant (otherwise, density perturbations are largest where $w'N'^2$ is largest). The velocity variations at this order are all related to these density changes through, for example, the thermal wind balance for alongshore velocity. The point is that including modest stratification does not change the problem radically, and the effects of stratification tend to be felt most strongly near the bottom in this strictly linear model. This contrasts strikingly with observations that tend to have the strongest density changes in the upper part of the water column (e.g., Figure 4.2). The contrast can be attributed to the perturbation problem's linearity and its neglect of surface mixed-layer processes.

## 4.5.   An Extreme Idealization

At very low latitudes, the deep-sea internal Rossby radius can become large (well more than 100 km) and so exceed the width of the shelf-slope topography. Thus, $Bu$ can be large over a limited latitude range. In this limit, it is possible to ignore the sloping bottom and treat the entire continental margin as a vertical wall, although the role of the inner shelf (chapter 6) becomes confused. With continuous stratification, the system (4.4.1) is still applicable, but the low-frequency assumption becomes problematic, because $f$ approaches zero. This problem with continuous stratification (and, importantly, its equatorial analog) was treated in three dimensions by Gill and Clarke (1974). For now, consider the two-dimensional problem as a two-layer, flat-bottom configuration (Figure 4.3) with the understanding that the results will mimic the behavior of the barotropic and first baroclinic modes (see section 5.4) of a more complete, continuously stratified solution. Although it is not relevant in the present flat-bottom context, two-layer models over a sloping bottom topography can give rise to peculiar results, so caution is required with this idealization (Chapman, 1984).

To proceed, the linearized equations of motion in a two-dimensional system with a rigid lid and no bottom friction, depth-integrated over each layer, are

$$-fV_1 = -gH_1\zeta_x, \tag{4.5.1a}$$

$$V_{1t} + fU_1 = \frac{1}{\rho_0}\tau_0^y, \tag{4.5.1b}$$

$$0 = U_{1x} - \eta_t, \tag{4.5.1c}$$

$$fV_2 = gH_2\zeta_x + g'H_2\eta_x, \tag{4.5.1d}$$

$$V_{2t} + fU_2 = 0, \tag{4.5.1e}$$

$$0 = U_{2x} + \eta_t, \tag{4.5.1f}$$

where $(U, V)$ represent depth-integrated velocities over a layer; subscripts $(1, 2)$ denote the upper and lower layers; $H_1$ and $H_2$ are the initial layer thicknesses; $(\zeta, \eta)$ are the displacements of the free surface and the interface, respectively (both small relative to $H_1$ and $H_2$); and $g'$ is the "reduced gravity," $g(\rho_2 - \rho_1)/\rho_2$. Although it is problematic for small $f$, a low-frequency assumption has also been made to eliminate near-inertial oscillations (section 5.12). The problem is to be solved subject to $U_1 = U_2 = 0$ at $x = 0$ and ensuring the solution is bounded far from shore.

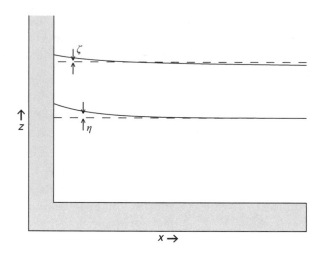

**FIGURE 4.3.** Schematic of a two-layer model of the wind response of a flat-bottom ocean.

If a steady wind forcing (4.3.4) is turned on abruptly, the solution for $t > 0$ becomes

$$U_1 = \frac{\left(\frac{\tau_A}{f\rho_0}\right)H_2(1-e^{-\gamma x})}{H_1+H_2}, \tag{4.5.2a}$$

$$V_1 = \left(\frac{\tau_A}{\rho_0}\right)t(H_1+H_2e^{-\gamma x})/(H_1+H_2), \tag{4.5.2b}$$

$$U_2 = \frac{-\left(\frac{\tau_A}{f\rho_0}\right)H_2(1-e^{-\gamma x})}{H_1+H_2} = -U_1, \tag{4.5.2c}$$

$$V_2 = \frac{\left(\frac{\tau_A}{\rho_0}\right)tH_2(1-e^{-\gamma x})}{H_1+H_2}, \tag{4.5.2d}$$

$$\eta = \left(\frac{\tau_A}{\rho_0}\right)\frac{ft}{g'\gamma H_1}e^{-\gamma x}, \tag{4.5.2e}$$

where $\gamma$ is the inverse of the internal Rossby radius:

$$\gamma^2 = \frac{f^2(H_1+H_2)}{g'H_1H_2}. \tag{4.5.2f}$$

There are some interesting aspects of this solution. With spatially uniform winds, the maximum density variation (interface displacement) is at the coast, while with bottom topography (section 4.4), the maximum can occur far from shore. Next, the cross-shelf flow ($U_1$, $U_2$) is perfectly steady, although the alongshore flow, hence vertical displacement $|\eta|$, constantly increases. The natural length scale of the problem is the internal

Rossby radius, and close to shore, there is a surface-intensified alongshore jet. Far from shore ($\gamma x \gg 1$), both the alongshore and cross-shelf (aside from the surface boundary-layer transport) velocity are depth independent; for example, $V_1/H_1 = V_2/H_2$.

The depth independence of flow far from shore makes sense in that, with a flat bottom, the only natural horizontal scales are the Rossby radii for the various available baroclinic modes (in this case, only the barotropic mode and the first baroclinic). The higher baroclinic modes all have relatively short length scales compared with the barotropic scale (which is infinite with a rigid lid), so that once a flat-bottom region is encountered, far enough offshore, flow can exist only in the barotropic mode: all the higher modes are confined closer to shore. Thus, the water drawn onshore compensating for an offshore surface Ekman transport of $\tau_A/(f\rho_0)$ ultimately has to come equally from all depths far enough from shore. The dominance of the barotropic mode far offshore occurs simply because the barotropic Rossby radius is much larger than the internal radius. This ultimate depth-independence holds as long as there is a flat-bottom region far offshore where the Rossby radius is well defined; the existence of shelf-slope topography or of continuous stratification makes no difference far offshore. This barotropic outcome flatly contradicts the observational evidence that upwelled water generally comes from the upper 100–200 m of the water column. Yet, this is a property common to all linear two-dimensional models as long as the wind stress is uniform in $x$. In fact, with bottom friction, where a steady state requires that surface and bottom stresses match, the barotropic dominance offshore requires that all the upwelled water comes from the abyssal bottom boundary layer. Thus, the unrealistic result that upwelled water eventually comes from abyssal depth is not relieved by stratification, but, rather, the deep source is required by the two-dimensional assumption. This difficulty becomes an issue over the time scale it takes for a water parcel to pass from the very deep offshore region onto the shallower shelf, and this advective time might be many tens of days for realistic conditions. Thus, on shorter time scales, two-dimensional models often appear quite realistic. When alongshore variability is allowed, this deep-source problem is alleviated because alongshore convergence/divergence in the upper few hundred meters can remove the requirement that source water must come from great depth far offshore.

## 4.6.    Nonlinear Effects

The linear models in the preceding sections elucidate how cross-shelf and vertical motions are established in response to an alongshore wind stress. But these models have their limits. For example, in the layer model, what happens as the interface shoals (so the upper-layer depth decreases nearshore) and $\gamma$ (4.5.2f) presumably becomes larger locally? Questions of this sort call for treating nonlinearities, hence, very often, the use of numerical models.

First, consider the case of coastal upwelling (e.g., Allen et al., 1995; Austin and Lentz, 2002). Both these two-dimensional numerical studies treated wind response in a continuously stratified ocean over realistic shelf-slope topography, and both used a spatially

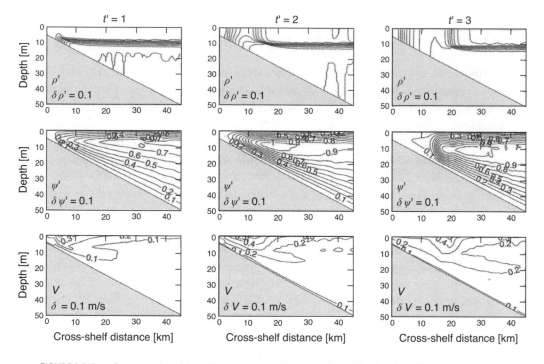

**FIGURE 4.4.** Two-dimensional model results demonstrating the onset of upwelling. Sections of density (upper row), (x, z) streamfunction (middle row), and alongshore velocity (lower row) demonstrating model response to upwelling-favorable wind stress. Different columns correspond to different nondimensional times. Adapted from Austin and Lentz (2002).

uniform positive wind stress. Although results were similar, there were some differences between the two studies, primarily because of differences in their U.S. west versus east coast summertime contexts. The gently sloping east coast (Austin and Lentz, 2002) case is featured here because the very sharp seasonal pycnocline makes for clearer illustrations.

After the onset of a wind stress, the seasonal pycnocline is rapidly brought to the surface and forms a density front there (Figure 4.4) which progressively moves offshore, carried by the surface Ekman transport. Below this front, deeper, denser water is carried shoreward. Because density below the pycnocline is relatively uniform in this model (but not in Allen et al., 1995), water inshore of the front is relatively homogeneous and weakly stratified. The exception to this condition occurs close to shore where the density is similar to the initial state. This occurs because in 5–10 m deep water (i.e., shallow relative to a turbulent boundary-layer thickness), conditions are highly dissipative, hence relatively stagnant, so there is little flushing from deeper offshore. Associated with the offshore-moving density front, through the thermal wind equation, is a well-defined alongshore jet which is consistently the strongest alongshore flow anywhere in the system. The example shown in Figure 4.4 begins with very strong stratification ($N^2 = 5 \times 10^{-3}$ 1/s$^2$) concentrated in the pycnocline, and so the surface mixed layer is very resistant to further deepening.

In the west coast case ($N^2 = 0.7 \times 10^{-3}$ 1/s$^2$ or less), this resistance is weaker, and the surface mixed layer offshore of the front deepens. This deepening is accelerated in models without surface heating by the gravitational instability associated with the surface Ekman layer advecting denser upwelled water over less dense ambient water. Interestingly, though, Lentz (1992) examined several sets of summertime observations in upwelling regions and showed that this tendency for upper-ocean cooling via offshore advection is balanced to a good approximation by the incoming surface heat flux, thus limiting mixed-layer deepening for his examples.

There are alternative ways to approach the upwelling problem. For example, Pedlosky (1978) considered only the inviscid interior of a continuously stratified ocean but allowed for density and momentum advection. In this case, a front forms below the surface mixed layer at the coastal boundary. This form of frontogenesis has distinct parallels in the atmospheric literature and can be thought of as a local runaway shrinkage of the natural scales (e.g., the internal Rossby radius) as the density field evolves. At another extreme, deSzoeke and Richman (1984) treated coastal upwelling as a two-dimensional mixed-layer model. This approach, although rather cumbersome, accounts for the difference in stratification and mixed-layer depth on either side of the front, as well as for the front's offshore translation.

When downwelling-favorable (negative) winds are applied, the response is far from symmetric (Figure 4.5). In this case, surface Ekman transport carries relatively light water onshore, and it occupies a growing area on the inner part of the shelf. These waters are very homogenous (compared with the upwelling case) because they all came from the same initial shallow depth, so there is little stratification to inhibit inner-shelf turbulence. Being highly turbulent, this region is again relatively stagnant in terms of lower-frequency flows. A downwelling front marks the location where bottom Ekman transport carries the relatively light inshore waters toward the pycnocline, resulting in a very sharply defined feature. Because the waters inshore of the front are essentially the same as surface waters offshore, the downwelling front has no surface hydrographic expression. This lack of contrast makes it difficult to detect the front using surface observations and may help explain why this feature is relatively infrequently studied in the field. Offshore of the downwelling front, the downslope bottom Ekman transport creates a thick bottom boundary layer (section 3.7), horizontal density gradients, and the potential for hydrodynamic instabilities (chapter 9). The downwelling front moves offshore more slowly than the corresponding upwelling front (Allen and Newberger, 1996; Austin and Lentz, 2002).

When the wind stress ceases, the alongshore jet adjusts to a geostrophically balanced state such that the alongshore current vanishes at the bottom (Austin and Lentz, 2002). This is true regardless of the sign of the wind stress. Thus, there is no bottom stress or Ekman transport, and the system does not evolve further. This means there is no inherent tendency for a two-dimensional front to flatten out or return toward the coast in the absence of wind forcing. This sort of relaxation to level isopycnals requires either a wind reversal or three-dimensional processes (coastal-trapped waves or instabilities), which are covered in chapters 5 and 9, respectively.

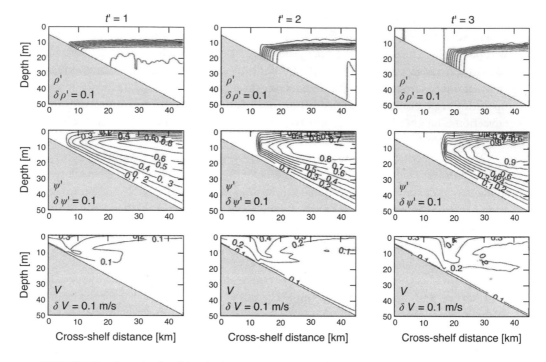

**FIGURE 4.5.** Two-dimensional model results demonstrating the onset of downwelling. Sections of density (upper row), $(x, z)$ streamfunction (middle row), and alongshore velocity (lower row). Different columns correspond to different nondimensional times. Adapted from Austin and Lentz (2002).

The presence of a baroclinic jet opens up new possibilities about the nature of the cross-shelf flow that balances the wind-driven Ekman transport offshore of the inner shelf (Lentz and Chapman, 2004). The alongshore momentum equation (4.2.1b with $p_y = 0$), including the advective cross-shelf transport of alongshore momentum, can be integrated over depth and from the coast out to some $x = x_0$ (where the water at $x_0$ is deep enough to have distinct surface and bottom boundary layers) to obtain

$$\int_0^{x_0} \int_{-h}^{0} v_t\, dz\; dx + \int_{-h}^{0} uv\, dz\big|_{x=x_0} = \frac{1}{\rho_0}\int_0^{x_0}(\tau_0^y - \tau_B^y)dx. \tag{4.6.1}$$

Referring to (4.3.2), each of these terms can be thought of as representing a cross-shelf transport (e.g., the surface and bottom Ekman transports on the right-hand side). Alternatively, terms can be thought of as growth, sources, or sinks of alongshore momentum. In the upwelling case (Figure 4.4), the alongshore flow is surface-intensified and in the direction of the wind, so that $v_z > 0$. Thus, close to the surface, the offshore Ekman transport ($u > 0$) gives rise to a dominating positive contribution to the $uv$ integral. At depth, either in the bottom Ekman layer or in the interior, $u$ is expected to be less than zero, but $v$ being weaker, this negative contribution to the integral cannot dominate. For upwelling, then, the advection term in (4.6.1) is positive—to balance the surface wind stress, at least partially.[2] So, near the frontal jet in the steady state, the wind stress need not be balanced by the bottom stress. Thus, in the presence of either a steady or time-varying alongshore flow, the bottom stress and bottom Ekman transport can weaken, so that proportionately more of the local onshore flow occurs above the bottom boundary layer. Indeed, Lentz and Chapman (2004) examined several observational data sets and found that this theory can account for otherwise unexplained midshelf $u(z)$ profiles. Further, they pointed out that similar reasoning also applies to the downwelling case.

## 4.7.    Conclusion

The two-dimensional models treated in this chapter provide a number of important insights, including the mechanisms for accelerating water-column alongshore flow, for forming and translating fronts, and for generating the asymmetry between upwelling and downwelling. In contrast with three-dimensional models, these simplified systems readily allow the fine resolution required for frontal numerical modeling. But these same models have distinct disadvantages, including the spatial uniformity of the alongshore pressure gradient, the requirement that at least some upwelled water eventually must come from great depth offshore, and, of course, the inability to address a range of inherently three-dimensional effects such as irregular topography or alongshore variations in wind strength. One aspect of this artificiality (rationalized in chapter 10) is that there is nothing in any of these two-dimensional models that inhibits cross-isobath flow, even though it is well known that on a rotating planet, velocities have a strong tendency to parallel depth contours.

   Indeed, two-dimensional models can leave the impression that the ocean is less complex than it actually is. In reality, most processes in the ocean are three-dimensional, and often strongly so. Smith (1981) used data to show, for example, that it is often impossible to find any $(x, y)$ coordinate system that allows a consistently two-dimensional cross-shelf flow, that is, one where the observed depth integral of $u$ is always nearly zero. Further, topographic features such as canyons (e.g., Allen and Durrieu de Madron, 2009), capes (Barth et al., 2000), and banks (Castelao and Barth, 2005) greatly complicate the response to alongshore winds. Detailed observations show that upwelled waters are often drawn offshore into narrow filaments as they pass through active eddy fields beyond the shelf break (e.g., Strub et al., 1991; Zaba et al., 2021). But in dealing with all these complexities, the two-dimensional model remains an essential starting point.

---

2 Similar reasoning helps explain why nearshore waters grow denser during upwelling. Shallow waters are lighter (warmer) than deeper waters, so warmer water is removed from the nearshore volume by the surface Ekman transport and is replaced by deeper, denser water.

# 5
# Waves in the Coastal Ocean

## 5.1. Introduction: Responses to Distant Forcings

Waves are the messengers of the ocean: they carry information and energy to locations far from where they are generated. Often, wave propagation is rapid compared with advection, which proceeds at the pace of the ocean currents. The coastal ocean acts as a particularly effective wave guide over a range of periods extending up to many days, so information can be carried efficiently, especially alongshore.

The response of sea level to hurricanes along the west coast of Mexico (Enfield and Allen, 1983; Figure 5.1) is a particularly clear example of waves in the coastal ocean. In this case, a hurricane on July 21–22, 1973, perturbed sea level at AC (Acapulco), and then a distinct 10–20 cm mound of water propagated northward past positions MN, MZ, TO, and GU at a rate of about 200–300 km/day. As a result, this hurricane affected the coastal ocean hundreds of kilometers north of where it encountered the coast. While this is a particularly clear example of free wave alongshore propagation, it is far from unique; similar alongshore propagation can be found off Australia, southwestern Africa, and the western United States, to cite just a few examples. The wave's physics will become clearer in this chapter, but the important point for now is the way information is carried alongshore so effectively.

There are several classes of waves in the coastal ocean. Some are specifically coastal, some are nearly the same as would be found elsewhere in the ocean, and some fall in between. For the broader oceanic context, a number of fine references exist, including LeBlond and Mysak (1978) and Pedlosky (2003). The focus here is on types of waves that are unique to the coastal ocean or where the coastal environment substantially affects the wave's behavior. This chapter deals only with hydrostatic waves, so that deep-water surface gravity waves (which, after transformation in shallow water, play such an important role in the surf zone) are treated separately in the context of the inner shelf (section 6.3). Some other wave classes, like high-frequency internal waves, do not receive much attention either, because their underlying dynamics are not much different from elsewhere in the ocean, even though their statistical properties (e.g., amplitude and propagation direction distributions) may differ from those elsewhere. Freely propagating (or

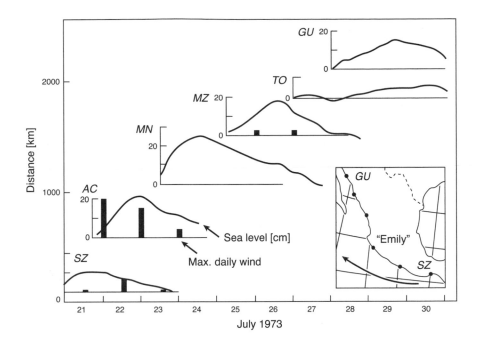

**FIGURE 5.1.** Sea level response to a coastal storm off the west coast of Mexico in 1973 (adapted from Enfield and Allen, 1983), demonstrating the northward propagation of a storm-induced sea level response. Each curve represents a fragment of a smoothed (tides and near-inertial waves removed) coastal sea level record, with the fragments chosen to demonstrate the response to Hurricane Emily. Letters represent different Mexican cities from Salina Cruz (SZ) in the south to Guaymas (GU) in the north. Each plot fragment is placed at its correct time, and axes are offset vertically to reflect alongshore separations. Heavy vertical bars represent wind magnitudes (m/s). The hurricane's path is shown on the inset map.

simply "free") waves as they behave in the absence of forcing are emphasized, but forced wave motions are also treated. The chapter starts with strong simplifications and relatively idealized results and then adds layers of realism to arrive at outcomes that include the combined effects of topography, stratification, and rotation.

## 5.2. An Unbounded, Rotating, Homogeneous, Flat-Bottom Ocean

The starting point is an unbounded ocean with a flat bottom, spatially constant rotation, and uniform density $\rho_0$. Further, only long (relative to water depth) waves are considered, so that pressure is in hydrostatic balance. In this case, the equations of motion for linear waves with no dissipation are, from (2.4.10),

$$u_t - fv = -\frac{1}{\rho_0} p_x = -g\zeta_x, \qquad (5.2.1a)$$

$$v_t + fu = -\frac{1}{\rho_0}p_y = -g\zeta_y, \tag{5.2.1b}$$

$$Hu_x + Hv_y + \zeta_t = 0. \tag{5.2.1c}$$

Subscripts with regard to independent variables $(x, y, t)$ represent partial differentiation, $(u, v)$ are horizontal velocity components, $p$ is pressure, $H$ is the constant mean water depth, $\zeta$ is the sea level perturbation (such that total water depth is $H + \zeta$ and $|\zeta| \ll H$), $f$ is the Coriolis parameter, and $g$ is the acceleration due to gravity. Using $p = g\rho_0 \zeta$, a single equation in terms of $p$ is readily found:

$$0 = \frac{1}{c_0^2}\left(\frac{\partial^2}{\partial t^2} + f^2\right)p - \frac{\partial^2 p}{\partial x^2} - \frac{\partial^2 p}{\partial y^2}, \tag{5.2.2a}$$

where

$$c_0 = \sqrt{gH} \tag{5.2.2b}$$

is the speed of long gravity waves in a nonrotating ocean. In the idealized case where only one sinusoidal wave is present, a solution is

$$p = G\sin(kx + ly - \omega t), \tag{5.2.3}$$

where $(k, l)$ are horizontal wavenumbers, and $\omega$ is frequency. Substituting (5.2.3) into (5.2.2a) yields the *dispersion relation* (i.e., the equation relating frequency to wavenumber)

$$\omega^2 = f^2 + c_0^2(k^2 + l^2). \tag{5.2.4}$$

This solution represents long gravity waves modified by Earth's rotation, and these are often called either inertia-gravity waves or *Poincaré waves*. One interesting property of this solution is that the frequency range is restricted. This type of wave is possible only when

$$\omega^2 \geq f^2, \tag{5.2.5}$$

and a valid solution can be found for any $(k, l)$ pair so long as either

$$k^2 \leq \frac{\omega^2 - f^2}{c_0^2} \text{ or } l^2 \leq \frac{\omega^2 - f^2}{c_0^2}, \tag{5.2.6}$$

that is, the shaded area in the dispersion plot, Figure 5.2. For very large wavenumbers (short scales), where $(k^2 + l^2)^{1/2}H$ reaches $O(1)$, this solution becomes invalid as the hydrostatic approximation begins to fail.

It is natural to ask how quickly these waves propagate, and the simplest answer lies in asking how fast one would need to move to always stay at, say, a peak of the sine function in (5.2.3) as it translates. This can happen if the phase is constant; that is, if

$$kx + ly - \omega t = -\omega t_0 \tag{5.2.7}$$

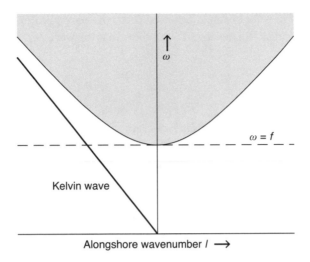

**FIGURE 5.2.** Dispersion diagram relating frequency $\omega$ to alongshore wavenumber $l$ for waves in a bounded, flat-bottom ocean with uniform density and $f > 0$. The shaded area represents the continuum of Poincaré waves that satisfy (5.2.4).

(where $t_0$ is a constant), a condition that requires

$$\frac{x}{t-t_0} = \frac{\omega k}{k^2 + l^2} = c_p^x, \tag{5.2.8a}$$

$$\frac{y}{t-t_0} = \frac{\omega l}{k^2 + l^2} = c_p^y. \tag{5.2.8b}$$

This quantity, the *phase velocity* $(c_p^x, c_p^y)$, describes the speed and direction with which wave crests or troughs move in a perfectly monochromatic wave that uniformly fills all space. It cannot describe energy propagation, because the energy is the same everywhere in space.

Energy propagation is described by a depth-integrated equation derived from (5.2.1):

$$0 = \frac{1}{2}[H(u^2 + v^2) + g\zeta^2]_t + gH(u\zeta)_x + gH(v\zeta)_y. \tag{5.2.9}$$

The wave energy term includes kinetic energy (the $u^2 + v^2$ component) and potential energy (the $g\zeta^2$ component). Changes in energy occur when there is a divergence in the energy flux, that is, when $\nabla \cdot \zeta(u, v) \neq 0$. It follows that if the energy and energy flux were perfectly uniform everywhere (divergence equal to zero), the energy would not change. Of course, there is always nonuniformity in nature: waves occur in groups or packets, so that, over time, the wave amplitude or direction at a set location varies. Thus, in reality, it is reasonable to expect the flux divergence to be nonzero and for wave energy to propagate from one place to another. Given some nonuniformity, the amplitude $G$ in (5.2.3) really ought to depend on space and time, albeit perhaps slowly relative to $k$, $l$, and $\omega$. Pedlosky (1979), for example, followed through the implications of using a variable $G$ in an equation analogous to (5.2.3) to show that the wave packet moves with the *group velocity*

$$(c_g^x, c_g^y) = \left( \frac{\partial \omega}{\partial k}, \frac{\partial \omega}{\partial l} \right). \tag{5.2.10}$$

This descriptor of the packet motion is in fact a very general property of most linear waves. A wave packet represents spatial/temporal variations in amplitude and energy, so that as the packet translates, the location of higher wave energy moves with it. It is thus intuitive that the packet, or group, velocity describes energy propagation. This supposition is readily confirmed by using the solution (5.2.3) along with (5.2.1) to obtain expressions for $u$ and $v$, then defining and averaging the various terms in (5.2.9) over a wave period. The averaged flux divergence terms then become

$$gH <u\zeta>_x = c_g^x \langle E \rangle_x, \tag{5.2.11a}$$

and

$$gH <v\zeta>_y = c_g^y \langle E \rangle_y, \tag{5.2.11b}$$

where angular brackets represent an average, and the averaged wave energy per unit mass is

$$<E> = \frac{1}{2} \langle H(u^2 + v^2) + g\zeta^2 \rangle. \tag{5.2.11c}$$

Again, the result that the energy transport can be represented in terms of the group velocity and wave energy is a very general one for linear waves.

A wave packet can always be Fourier decomposed and so represented as a collection of perfectly sinusoidal waves having a range of wavenumbers. For Poincaré waves, the phase and group velocities both depend on the wavenumber, so that different Fourier components of a wave packet move at different speeds, eventually leading to the packet spreading out in space and separating by frequency/wavelength. That is to say that the packet disperses into a range of sinusoidal waves over time. This is true even if the initial packet is simply a mound of water that does not appear at all sinusoidal. Thus, waves whose speed varies with wavelength are called *dispersive*. Waves whose speed does not vary with wavelength, that is, whose frequency depends strictly linearly on wavenumber, are *nondispersive* and propagate without changing the initial packet shape. To return to Poincaré waves, they are dispersive in general, but for frequencies high relative to the Coriolis parameter (so that the $f^2$ term in (5.2.4) is negligible) they become nondispersive.

Poincaré waves provide an introduction to wave dispersion, phase propagation, and energy propagation. Similar discussions and conclusions could be made for all the other classes of waves discussed in this chapter. Thus, the definitions (5.2.8) and (5.2.10) will be applied here to linear ocean waves in general.

## 5.3.  A Rotating, Homogeneous, Flat-Bottom Ocean with a Coastal Wall

Next, add a vertical coastal wall at $x=0$, and consider waves in the constant-depth ocean, $x>0$. The governing equations are, as before, (5.2.1) and (5.2.2). The difference is that now there is no flow across the coastal boundary; that is, $u=0$, or, from (5.2.1),

$$0 = \zeta_{xt} + f\zeta_y \qquad \text{at } x=0. \tag{5.3.1}$$

Poincaré wave solutions exist in this case, but only as reflecting pairs so that the boundary condition is satisfied. In addition, there is a wave that propagates strictly alongshore and that can have the form

$$\zeta = Z(x)\,\sin(ly - \omega t), \tag{5.3.2}$$

so that (5.2.2) becomes

$$0 = Z_{xx} - l^2 Z - \frac{(f^2 - \omega^2)}{gH} Z, \tag{5.3.3}$$

and (5.3.1) becomes

$$0 = -\omega Z_x + lZ, \tag{5.3.4a}$$

or

$$0 = \frac{Z_x}{Z} - \frac{lf}{\omega} \qquad \text{at } x = 0. \tag{5.3.4b}$$

A solution to (5.3.3) that is well behaved (bounded) far from shore is

$$Z = Be^{-x/\lambda}, \tag{5.3.5}$$

where (from 5.3.3) $\lambda$ is the positive solution of

$$\lambda^{-2} = l^2 + \frac{(f^2 - \omega^2)}{c_0^2}, \tag{5.3.6}$$

(where $c_0$ is defined in eqn. 5.2.2b), and from the boundary condition (5.3.4),

$$\lambda^{-1} = -\frac{fl}{\omega}. \tag{5.3.7}$$

Combining these leads to the simple outcome that

$$\lambda = c_0/f, \tag{5.3.8}$$

and

$$\omega = -c_0 l. \tag{5.3.9}$$

Further, when (5.3.5) and (5.3.7) are used in (5.2.1), it turns out that $u=0$ everywhere. This is a consequence of the form (5.3.5) requiring that $Z_x/Z = -\lambda$, a constant. In addition, from (5.2.1a), the alongshore flow is in perfect geostrophic balance, because $u = 0$. The cross-shelf scale of the wave, $\lambda$ (5.3.8), is known as the *Rossby radius of deformation*, and it appears in most problems involving Earth's rotation and divergent flow. It can be thought of as the distance that a long gravity wave travels in a time $f^{-1}$.

This exponentially decaying solution is the *barotropic Kelvin wave* (denoted by the heavy line in Figure 5.2), and it is really a remarkable result. The alongshore momentum balance is that of a long gravity wave in a nonrotating system even at subinertial ($\omega < f$) frequencies, but the alongshore flow is in perfectly geostrophic balance even at superinertial ($\omega > f$) frequencies. That alongshore propagation proceeds at the nonrotating speed $c_0$ follows from (5.2.1b, c) where rotation drops out when $u = 0$. Additionally, alongshore propagation (from 5.3.9) can only go toward $-y$ in the northern hemisphere (a consequence of having $Z_x / Z < 0$, as required for offshore decay, in 5.3.4b). This one-way propagation is typical of the anisotropy that permeates coastal dynamics on time scales longer than $f^{-1}$. Finally, this solution exists at any frequency, no matter how high, so long as conditions remain hydrostatic.

The cross-shelf scale $\lambda$ of this barotropic Kelvin wave, in practice, is rather large: $O(2000$ km) in the open ocean at midlatitudes. Thus, the natural scale is much greater than the 100 km typical width of a continental shelf/slope. The upshot is that on the oceanic barotropic Rossby radius scale, any coastal topography is narrow, and the continental margin is a vertical wall to a good first approximation (Smith, 1972). The effect of the shelf-slope topography is greater when the Rossby radius is smaller relative to the topographic width—a condition that might occur in a shallow lake, for example.

Near the equator, the Coriolis parameter $f$ cannot be treated as a constant but, rather, $f \approx \beta y$, where $y$ is a northward coordinate. In this case, there is also an equatorial Kelvin wave, trapped near the equator, that propagates strictly eastward at speed $c_0$ (e.g., Clarke, 2008; section 10.2).

Kelvin waves play a major role in basin-scale tidal currents (see chapter 7). Tide-generating forces excite motions in the open ocean, and typically the response is associated with a range of basin modes that are excited to differing degrees, depending on the match of frequency and basin geometry to the forcing. Among these basin modes are many that are dynamically similar to barotropic Kelvin waves. Global tidal syntheses (e.g., Egbert et al., 1994) thus exhibit many patterns that resemble Kelvin waves. For example, in both the North Pacific and North Atlantic, there are large-scale pressure disturbances that are strongest at the coast and propagate phase counterclockwise around the basin at the sort of speeds, $O(200$ m/s), expected for a Kelvin wave.

## 5.4.    Including Density Stratification with a Flat Bottom

It is now remarkably straightforward to include the effects of vertical density stratification if the background density field varies only in the vertical. The flat bottom and constant rotation assumptions continue to hold for now. The linearized hydrostatic governing equations are then

$$u_t - fv = -\frac{1}{\rho_0} p_x, \tag{5.4.1a}$$

$$v_t + fu = -\frac{1}{\rho_0} p_y, \tag{5.4.1b}$$

$$0 = -p_z - g\rho_2, \tag{5.4.1c}$$

$$0 = \rho_{2t} + w\,\rho_{1z}, \tag{5.4.1d}$$

$$0 = u_x + v_y + w_z, \tag{5.4.1e}$$

and the density is broken into

$$\rho_0 + \rho_1(z) + \rho_2\,(x,\,y,\,z,\,t), \tag{5.4.2a}$$

where

$$\rho_0 \gg |\rho_1| \gg |\rho_2|. \tag{5.4.2b}$$

It is then straightforward to reduce the set (5.4.1) to a form analogous to the barotropic equation (5.2.2a)

$$0 = p_{xxt} + p_{yyt} + \left( f^2 + \frac{\partial^2}{\partial t^2} \right)\left( \frac{p_{zt}}{N^2} \right)_z, \tag{5.4.3}$$

with the buoyancy frequency squared defined by

$$N^2 = -\frac{g\rho_{1z}}{\rho_0}. \tag{5.4.4}$$

Equation (5.4.3) represents a balance between the change in relative vorticity $v_x - u_y$ (i.e., the terms with horizontal derivatives in 5.4.3) and the vortex stretching associated with changes in vertical separation between isopycnals. Once again, this problem is solved for no flow through the coast ($u = 0$),

$$0 = fp_y + p_{xt} \qquad \text{at } x = 0, \tag{5.4.5}$$

and boundedness far from shore. No flow passes through the bottom,

$$0 = p_{zt} \qquad \text{at } z = -H, \tag{5.4.6a}$$

(from combining eqns. 5.4.1c and d), and the free-surface condition is that $w = \zeta_t$, where $\zeta$ is the free-surface perturbation, so that

$$0 = \frac{N^2}{g} p_t + p_{zt} \qquad \text{at } z = 0. \tag{5.4.6b}$$

Because the time derivative appears everywhere in (5.4.3) and in (5.4.6), these can be integrated once in time, so that this time derivative vanishes in the following.

At this point, one can apply the separation of variables to obtain

$$p = P(x,\,y,\,t)F(z), \tag{5.4.7}$$

so that (5.4.3) becomes

$$0 = F(P_{xx} + P_{yy}) + \left( f^2 + \frac{\partial^2}{\partial t^2} \right) P \left( \frac{F_z}{N^2} \right)_z, \tag{5.4.8}$$

and

$$\frac{P_{xx} + P_{yy}}{\left( f^2 + \dfrac{\partial^2}{\partial t^2} \right) P} = \frac{-\left( \dfrac{F_z}{N^2} \right)_z}{F} = \frac{1}{c^2}. \tag{5.4.9}$$

The boundary conditions (5.4.5–5.4.6) are also consistent with the separation (5.4.7). Thus, a function of only $z$ in (5.4.9) equals a function of only $(x, y, t)$, so they both must equal the same unknown constant, taken to be $1/c^2$. (The sign choice and square reflect some a priori knowledge of the answer.)

The vertical structure problem then becomes

$$0 = \frac{1}{c^2} F + \left( \frac{F_z}{N^2} \right)_z, \tag{5.4.10a}$$

subject to (from 5.4.6a)

$$0 = F_z \qquad \text{at } z = -H \tag{5.4.10b}$$

and (from 5.4.6b)

$$0 = \frac{N^2}{g} F + F_z \qquad \text{at } z = 0. \tag{5.4.10c}$$

Taken together, this is an *eigenvalue* problem that has an infinite number of solutions, given well-behaved buoyancy frequency profiles. Thus, henceforth the solutions are subscripted with a mode number $n$; that is, $(F_n, c_n)$ is an eigenfunction-eigenvalue pair. The eigenfunctions are orthogonal according to

$$\delta_{nm} = \int_{-H}^{0} F_n F_m dz, \tag{5.4.11}$$

where $\delta_{nm}$ is the Kronecker delta ($= 1$ for $n = m$, and $= 0$ otherwise). As an example, if $N^2 = N_0^2$, a constant, then

$$F_0 \cong a_0, \ c_0 = \sqrt{gH} \tag{5.4.12a}$$

$$F_n \cong a_n \cos\left( \frac{n\pi z}{H} \right), \ c_n = \frac{N_0 H}{n\pi} \qquad \text{for } n \geq 1, \tag{5.4.12b}$$

with the coefficients $a_n$ chosen so that (5.4.11) holds. (Equations 5.4.12 actually reflect an approximation, but a very good one, that $c_0$ is large.) For more realistic density profiles, it is straightforward to solve (5.4.10) numerically: again, there is a set of modes, each one having one more zero crossing than the mode before it (e.g., Figure 5.3). Collectively, the

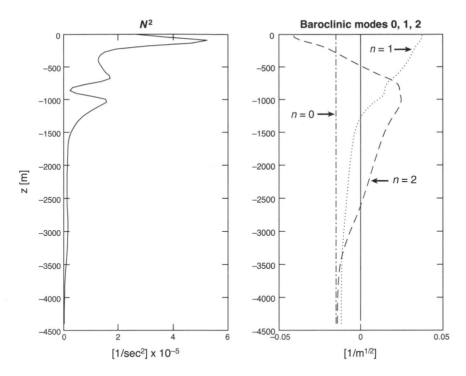

**FIGURE 5.3.** Representative midocean (near Bermuda) profiles of buoyancy frequency squared (left panel) and of the first three baroclinic modes $F_n$ (right panel). The wave speeds associated with each mode are $(c_0, c_1, c_2) = (208, 2.8, 1.3)$ m/s.

$F_n$ are called the *baroclinic modes*, but $F_0$, which is part of this collection, is referred to as the *barotropic mode*. Also, solutions for modes $n \geq 1$ are sometimes called *internal modes*.

Now, for each value of $n$, the $P$ part of (5.4.9) becomes

$$0 = P_{nxx} + P_{nyy} - \left( f^2 + \frac{\partial^2}{\partial t^2} \right) \frac{P_n}{c_n^2}, \tag{5.4.13}$$

subject to

$$0 = fP_{ny} + P_{nxt} \qquad \text{at } x = 0 \tag{5.4.14}$$

and boundedness far from shore. Note that, aside from the mode number, (5.4.13) is exactly the same as (5.2.2) and that (5.4.14) is similarly identical to (5.3.1). Thus, the solutions are the same: for each mode $n$, there is a Poincaré continuum (consisting of reflected wave pairs), along with a Kelvin wave. The only difference is the substitution of $n$ for 0 in the subscripts. The decreasing $c_n$ with increasing $n$ however means that each succeeding Kelvin wave is slower and that each Poincaré continuum covers a wider range in $l$. Likewise, in addition to the barotropic radius of deformation (5.3.8), there is now a set of internal Rossby radii

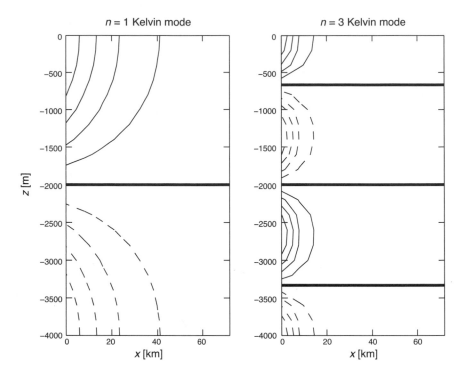

**FIGURE 5.4.** Changes in baroclinic modal structure as mode number increases: pressure structures as a function of $x$ and $z$ for an $n = 1$ internal Kelvin wave (left) and an $n = 3$ wave (right). Computed using $N_0^2 = 4 \times 10^{-6}$ 1/s$^2$ and $f = 1 \times 10^{-4}$ 1/s, so that $c_1 = 2.55$ m/s and $c_3 = 0.86$ m/s. Heavy lines are 0 contours, other solid contours are positive, and dashed contours are negative.

$$\lambda_n = \frac{c_n}{f}, \qquad\qquad\qquad (5.4.15)$$

so that as the mode number increases, the vertical "wiggliness" increases, and the horizontal length scale grows shorter in conjunction with the slower internal wave speed. This is illustrated in Figure 5.4: an example with constant $N^2$, so that the $F_n$ are described by (5.4.12). For reference, the mode number is the same as the number of zero crossings in pressure along the vertical coastal wall.

In conclusion, including stratification leads to shorter length scales (both vertically and offshore), as well as slower wave propagation. One troublesome aspect of this finding regards shelf-slope topography. For barotropic Kelvin waves (section 5.3), the Rossby radius is huge, about 2000 km at midlatitudes, so it is not hard to imagine that shelf-slope topography, having a width of perhaps 100 km, plays a minor role. With the first deep-water baroclinic mode, however, the lowest mode internal Rossby radius might be 25–50 km at midlatitudes, a width comparable to the topographic scale. This calls into doubt whether the vertical coastal wall is a reasonable approximation in a stratified ocean outside of the lowest latitudes. This concern helps motivate consideration of the effects of shelf-slope topography in the next section.

## 5.5.    Barotropic Waves over Shelf-Slope Topography

It is useful to consider what sorts of barotropic wave motions are possible over topography before dealing with the combined effects of stratification and topography. The starting point is the linearized hydrostatic equations of motion with no dissipation in a system where the water depth $h(x)$ now varies only offshore ($h$ represents a variable depth, and $H$ represents a constant depth). Thus, the depth-averaged linear hydrostatic frictionless equations of motion become

$$u_t - fv = -\frac{1}{\rho_0} p_x = -g\zeta_x, \qquad (5.5.1a)$$

$$v_t + fu = -\frac{1}{\rho_0} p_y = -g\zeta_y, \qquad (5.5.1b)$$

$$(hu)_x + hv_y + \zeta_t, \qquad (5.5.1c)$$

subject to coastal boundary condition (5.3.1). Taking

$$p = P(x) \sin (ly - \omega t), \qquad (5.5.2)$$

this set reduces to

$$0 = (hP_x)_x - \left[ \frac{flh_x}{\omega} + l^2 h + \frac{f^2 - \omega^2}{g} \right] P. \qquad (5.5.3)$$

This form (accounting for 5.5.2) can be compared with (5.2.2a) or (5.4.3): the new development here is the $flh_x$ term, which expresses changes in vorticity associated with cross-isobath flow and its consequent vortex stretching. Once again, formulation (5.5.3) allows for a Poincaré wave continuum (Figure 5.2) whose boundary is set by the fastest available gravity wave, $c_0^2 = gH$, with $H$ being the uniform depth of the adjoining deep ocean. Further, there is a mode with no zero crossings in the pressure field, and this will be seen to be the barotropic Kelvin wave, although it is slightly modified by the topography for $\omega < f$, and for $\omega > f$ it is increasingly confined to the shelf itself.

Now, however, two new sorts of waves appear.

### Edge Waves

First are the edge waves, represented by the discrete dispersion curves with $\omega > f$ in Figure 5.5. Although these waves were known for some time, they were relatively unappreciated (e.g., Lamb, 1932) until the second half of the twentieth century. As will be seen, edge waves are essentially shallow-water gravity waves that are confined to the inner part of the shelf because propagation is inhibited in deeper water.

The physics of these waves can be readily appreciated by considering a step-shelf geometry:

$$h = H_1 \qquad \text{for } x < L, \tag{5.5.4a}$$

$$h = H_0 \qquad \text{for } x > L, \tag{5.5.4b}$$

where $H_1 < H_0$ and there is no rotation ($f = 0$). Solving (5.5.3) with a wall at $x = 0$ leads to a trapped-wave solution:

$$P = C_1 \cos[\kappa(x-L)] + C_2 \sin[\kappa(x-L)] \qquad \text{for } x < L, \tag{5.5.5a}$$

$$P = G\, e^{-\gamma(x-L)} \qquad \text{for } x > L, \tag{5.5.5b}$$

when

$$\frac{\omega^2}{gH_0} < l^2 < \frac{\omega^2}{gH_1}, \tag{5.5.5c}$$

with

$$\gamma^2 = l^2 - \frac{\omega^2}{gH_0}, \tag{5.5.5d}$$

$$\kappa^2 = \frac{\omega^2}{gH_1} - l^2. \tag{5.5.5e}$$

The inequalities in (5.5.5c) express the notion that given $\omega$ and $l$, Poincaré waves are possible in the shallower water but not the deeper. Applying boundary conditions of no flow through the coast, and matching pressure and offshore transport at $x = L$, leads to

$$\frac{H_0}{H_1}\gamma = \kappa \tan(\kappa L). \tag{5.5.5f}$$

From the periodic nature of this relation, it follows that multiple solutions are possible, corresponding to a different number of zero crossings in pressure over the shelf (i.e., where $x < L$). When $l^2 < \dfrac{\omega^2}{gH_0}$, in contrast with condition (5.5.5c), Poincaré waves can exist for all $x$, and there are no trapped solutions. The important point about this example lies not in the details but in (5.5.5c): trapped waves, which can propagate in either alongshore direction, exist in the limited range where Poincaré waves are possible in the shallower but not in the deeper water.

A more realistic simple case occurs with an unbounded "beach" topography, so that $h = \alpha x$ (e.g., Reid, 1958). Again assuming that $f$ is negligible, there is an infinite set of solutions to (5.5.3) in the form

$$P(x) = \frac{e^{-|l|x}}{n!} L_n(2|l|x), \tag{5.5.6a}$$

with

$$\omega^2 = (2n+1)g\alpha|l|, \tag{5.5.6b}$$

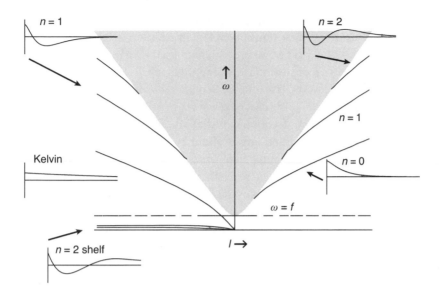

**FIGURE 5.5.** A schematic dispersion diagram (wave frequency $\omega$ as a function of alongshore wavenumber $l$) displaying the sorts of wave modes that appear in a barotropic ocean with shelf-slope topography and $f > 0$. The shaded area represents the Poincaré continuum. Curves for $\omega > f$ represent either edge waves or the Kelvin wave. The curves for $\omega < f$ are either shelf waves or the Kelvin wave. Examples of modal pressure structures are also shown.

and the $L_n$ are Laguerre polynomials (Abramowitz and Stegun, 1965); for example,

$$L_0(q) = 1, \tag{5.5.6c}$$

$$L_1(q) = 1 - q, \tag{5.5.6d}$$

$$L_2(q) = \frac{1}{2}\left(1 - 2q + \frac{1}{2}q^2\right), \dots \tag{5.5.6e}$$

(An implicit assumption is made that the wave decays offshore quickly enough that the hydrostatic approximation remains valid.) There are a few important aspects of this solution. One is that the waves propagate alongshore in both directions, and they are highly dispersive (5.5.6b). Second, the solutions all die off exponentially with distance from the coast (5.5.6a), and the decay is more rapid with larger $l$ or $\omega$. There is an infinite set of solutions, each having a higher frequency (5.5.6b) and each having more zero crossings with distance from shore (5.5.6c–e). Publicly available software for finding solutions with more realistic geometries and including mean flows can be found at https://darchive .mblwhoilibrary.org/handle/1912/26311.

Rotation ($f > 0$) leads to a slight asymmetry between positive and negative $l$ for modes $n \geq 1$. The edge wave mode having no zero crossing in $x$ ($n = 0$) and that propagates in the $-y$ direction (positive $\omega$ and negative $l$ in Figure 5.5) is special. In this case, rather than having the dispersion curve taper into the edge of the Poincaré continuum, the mode is continuous with the subinertial barotropic Kelvin wave. Following along the dispersion curve from small to larger $|l|$, the modal structure contracts from having an extremely large offshore scale (relative to the topographic width), $c_0/f$, when $\omega \leq f$, to being confined to the inner kilometer of the shelf for $\omega \gg f$. These transitional Kelvin-edge waves have been detected as a dramatic response to a hurricane passage, although their dispersion properties become more complicated when a more realistic topography is applied (Yankovsky, 2009).

High-frequency edge waves have often been observed along plane sandy beaches (e.g., Oltman-Shay and Guza, 1987). The observed waves are driven by momentum fluxes carried onshore by gravity waves (section 6.3) that are, in turn, generated by the wind. Observed edge waves often have periods of about 20–100 s, alongshore wavelengths of 50–500 m, and cross-shelf scales of 100 m or less (so that water depth is typically $O(5\text{ m})$ or less)—scales that fall in a range where Earth's rotation is not relevant. Because of the scales involved, it is possible to deploy intensive instrument arrays that allow the calculation of observed frequency-wavenumber spectra that, in turn, can be compared with theoretical dispersion curves (e.g., Oltman-Shay and Guza, 1987; Figure 9.9 of this volume), an impressive observational achievement.

## Continental Shelf Waves

The second new class of waves, barotropic continental shelf waves, exists only for $\omega < |f|$ and depends critically on Earth's rotation for its existence (see the curves with $\omega < f$ in Figure 5.5). While edge waves express a continuing exchange between potential energy (surface elevation) and kinetic energy, shelf waves express a continuing interplay of relative vorticity (represented by $P_{xx} - l^2 P$ in eqn. 5.5.3) and vortex stretching [expressed through the cross-isobath flow term $(f\,l/\omega)h_x P$]. This dependence on Earth's rotation introduces substantial asymmetries to the problem, including the requirement that phase propagates only toward $-y$ in the northern hemisphere (assuming that depth increases monotonically offshore; Huthnance, 1975).

Because shelf waves depend on a sloping bottom, their structure is concentrated over the shelf and slope rather than in the flat-bottom deep ocean offshore. Hence, a representative cross-shelf scale for a shelf wave is $L$, the width of the shelf-slope topography, which is typically $O(30\text{–}100\text{ km})$. Assuming that $\omega$ is not large compared with $f$, the surface divergence term in (5.5.3) (the term including $g$) thus compares with the relative vorticity term as

$$D = \frac{f^2 L^2}{gH};$$

(5.5.7)

that is, with the squared ratio of the topographic width to the barotropic Rossby radius (typically $O(2000\text{ km})$ in the open ocean), which is quite small for most shelf topographies. For this reason, analytical theories involving shelf waves very often use the rigid-lid approximation, hence the convenient deletion of the divergence term in (5.5.3). The assumption that $D$ is negligible is equivalent to saying that the barotropic gravity wave speed is infinite. Some caution is required on this point, though, in the case of broad, flat shelves. In this situation, the local barotropic Rossby radius scale over a relatively shallow, $O(100 \text{ m})$ deep, shelf can be more like 300 km, hence more comparable to the local topographic scale. Even in this case, however, making the rigid-lid approximation does not usually create major errors in estimating shelf wave properties.

It is useful to explore the properties of shelf waves using an analytical example with a rigid lid and the topography (see the inset in Figure 5.6)

$$h = h_0 \, e^{2\lambda x} \qquad \text{for } x < L, \tag{5.5.8a}$$

$$h = h_0 \, e^{2\lambda L} \qquad \text{for } x > L. \tag{5.5.8b}$$

Solutions require that there be no flow through the coast, matching pressure and $u$ at $x = L$, and boundedness far from shore. Under these conditions, the solution to the rigid-lid version of (5.5.3) is

$$P(x) = C_1 e^{-\lambda x} \sin(\mu_n x) + C_2 e^{-\lambda x} \cos(\mu_n x) \qquad \text{for } x < L, \tag{5.5.9a}$$

$$P(x) = E e^{-|l|(x-L)} \qquad \text{for } x > L, \tag{5.5.9b}$$

where

$$\mu_n^2 = -\frac{2 f \lambda l}{\omega} - l^2 - \lambda^2. \tag{5.5.10}$$

Applying the boundary conditions (no flow through the coast, and matching pressure and $u$ at $x = L$) to eliminate $C_1$, $C_2$, and $E$ leads to

$$0 = \tan(\mu_n L) + \mu_n (\lambda - l)^{-1}. \tag{5.5.11}$$

Plotting out the two terms in (5.5.11) and remembering that $l < 0$ shows that there are an infinite number of $\mu_n$ solutions (Figure 5.6), with each lying in the range of

$$\left(n - \frac{1}{2}\right)\pi < \mu_n L < n\pi, \tag{5.5.12}$$

where $n$ is a positive integer. Once $\mu_n$ is known, the dispersion relation is then found from (5.5.10):

$$\omega = \frac{-2 f \lambda l}{\mu_n^2 + l^2 + \lambda^2}, \tag{5.5.13}$$

which has a maximum frequency (so the group velocity $\partial \omega / \partial l = 0$), where

$$l^2 = \mu_n^2 + \lambda^2. \tag{5.5.14}$$

Note that both $\mu_n$ and $\lambda$ scale roughly as $L^{-1}$. It is clear from (5.5.13) that wave frequency decreases as $\mu_n$ increases (i.e., as mode number and thus number of zero crossings of $P$ increase). The form of the dispersion relation (5.5.13) serves as a reminder that these are essentially topographic Rossby waves that have many properties in common with planetary waves on a $\beta$-plane (e.g., Pedlosky, 2003), although they are quantized because of the finite width of the sloping-bottom waveguide.

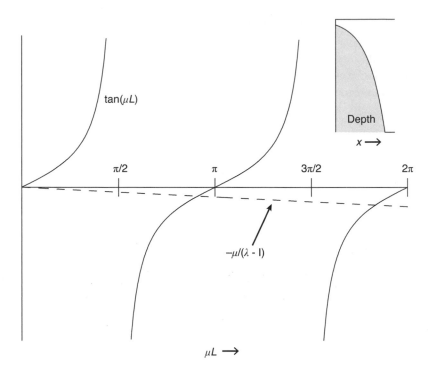

**FIGURE 5.6.** Plot of the two terms in (5.5.11). Solutions for $\mu$ occur where the two curves intersect. The inset shows the topography used for this calculation as a function of distance offshore.

For small $|l|$ (i.e., for large alongshore scales, $|lL| \ll 1$), $l^2$ is negligible, and the dispersion relation (5.5.13) is linear; that is, the waves are nondispersive (so that each phase velocity $\omega/l$ equals the corresponding group velocity $\partial\omega/\partial l$), and higher modes propagate more slowly. For large $|l|$ ($|lL| \gg 1$), the wave frequency with this geometry is proportional to $-l^{-1}$, so that $\partial\omega/\partial l > 0$; that is, the waves propagate energy in the direction opposite to the phase speed. Although these properties of the dispersion curve are demonstrated here for a simple analytical example, the results are actually rather general for the barotropic problem as long as $h_x/h$ is well behaved everywhere (Huthnance, 1975). This dispersion behavior is intuitive in that because shelf waves cannot exist for $\omega > |f|$, the frequency cannot indefinitely increase linearly with negative wavenumber, so the dispersion curves must "bend over" so as not to reach $f$. The software cited earlier for calculating edge wave modes under realistic conditions also works for barotropic shelf waves.

## 5.6.    Coastal-Trapped Waves with Stratification and Shelf-Slope Topography

In the actual coastal ocean, there are rarely situations in which the effects of bottom topography are negligible, and likewise, density stratification is very often important except for the cases of edge waves or barotropic Kelvin waves. It is particularly helpful to have some criterion to distinguish when such approximations are valid. Examining the physics of the more general case with both topography and stratification clarifies these questions and represents a further step toward realistic models.

It is useful to clarify some nomenclature. "Coastal-trapped wave" (or "hybrid coastal-trapped wave") is usually taken in the literature to describe a wave model that includes the effects of both stratification and bottom topography. Baroclinic (internal) Kelvin waves are then the end member where the coastal boundary is effectively a vertical wall, and barotropic shelf waves are the opposite extreme, where stratification can be ignored. This nomenclature is less than ideal, however, since some waves that are coastally trapped, such as edge waves, are left out. In practice, this does not seem to be a concern, because edge waves are often treated in a separate, higher-frequency, context.

In the presence of topography, the interior vorticity balance with linear physics is the same as it would be over a flat bottom (5.4.3):

$$0 = p_{xx} + p_{yy} + \left( f^2 + \frac{\partial^2}{\partial t^2} \right) \left( \frac{p_z}{N^2} \right)_z , \tag{5.6.1}$$

and boundary conditions at the coast and the free surface are the same as in the flat-bottom case:

$$0 = fp_y + p_{xt} \qquad \text{at } x = 0 , \tag{5.6.2}$$

$$0 = \frac{N^2}{g} p + p_z \qquad \text{at } z = 0. \tag{5.6.3}$$

What is new is the revised condition of no flow through the bottom:

$$0 = w + h_x u \qquad \text{at } z = -h(x), \tag{5.6.4}$$

or, in terms of pressure only,

$$0 = \left( f^2 + \frac{\partial^2}{\partial t^2} \right) \frac{p_{zt}}{N^2} + h_x (p_{xt} + fp_y) \qquad \text{at } z = -h(x). \tag{5.6.5}$$

This condition links the vertical and horizontal flow components, and effectively forces the interior with a vertical velocity at the bottom. The bottom constraint (5.6.5) and its location also make it difficult to find analytical solutions except in the very simplest of cases, but there is publicly available software for numerically finding the subinertial trapped-wave solutions of (5.6.1–5.6.5) under fairly general circumstances, namely, https://darchive.mblwhoilibrary.org/handle/1912/10527.

Not surprisingly, the wave properties reflect a blend of baroclinic Kelvin wave (section 5.4) and barotropic shelf wave (section 5.5) properties. The solutions were treated thoroughly by Huthnance (1978). For example, with well-behaved bottom topography, there is an infinite set of subinertial coastal-trapped waves, all with phase propagation toward the $-y$ direction in the northern hemisphere. Higher modes have increasingly complex cross-shelf and/or vertical structures and propagate more slowly.

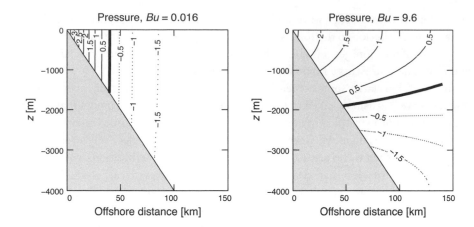

**FIGURE 5.7.** Coastal-trapped wave pressure modal structures for similar calculations having large alongshore wavelengths but different density stratification. The heavy contour represents zero pressure. Left panel: Burger number $Bu = 0.016$ ($N^2 = 1 \times 10^{-7}$ 1/s$^2$), $\omega = 2.76 \times 10^{-7}$ 1/s. Right panel: $Bu = 9.6$ ($N^2 = 6 \times 10^{-5}$ 1/s$^2$), $\omega = 10.1 \times 10^{-7}$ 1/s. In both cases, $l = -1 \times 10^{-7}$ 1/m, $f = 1 \times 10^{-4}$ 1/s.

However, to grasp the interplay of internal Kelvin versus shelf wave physics, it is useful to scale (5.6.1) so that time is nondimensionalized by $f^{-1}$, horizontal distance by $L^x$ (the width of the combined shelf and slope), vertical distance by $H$ (the deep-ocean depth), and $N^2$ by a representative buoyancy frequency squared $N_0^2$. This scaling leads to an important nondimensional parameter (see also eqn. 4.4.6), the Burger number

$$Bu = \frac{N_0^2 H^2}{f^2 L^{x2}}. \tag{5.6.6}$$

When $Bu$ is small, (5.6.1) reduces to a statement (see eqn. 4.4.6) that $w_z$ is small, implying barotropic dynamics governed by the bottom boundary condition. When $Bu$ is $O(1)$ or larger, relative vorticity changes in (5.6.1) are also associated with vertical stretching within the water column; that is, baroclinic dynamics apply. One can interpret $Bu$ as the squared ratio of the $n = 1$ flat-bottom baroclinic mode's Rossby radius (magnitude $N_0 H/f$) to the topographic width. Thus, if the deep-ocean internal Rossby radius is small compared with the topographic width, barotropic dynamics apply. Alternatively, using the approximation that shelf wave speed goes as $c = O(f L^x)$ (by scaling 5.5.13, or see section 5.7), (5.6.6) can be thought of as the squared ratio of the pure lowest-mode internal Kelvin wave speed $N_0 H$ to that of the gravest barotropic shelf wave. Or, yet again, (5.6.6) can also be thought of in terms of the deformation scale (4.4.5: the natural vertical scale obtained from scaling eqn. 5.6.1 and assuming small $\omega/f$),

$$L^z = (f/N_0)L^x. \tag{5.6.7}$$

If $L^z$ is large compared with $H$, then baroclinic effects occur on a vertical scale that is large relative to the water depth, and so the flow appears barotropic. If $L^z < H$, solutions tend to be trapped near the bottom; that is, the structure is baroclinic.

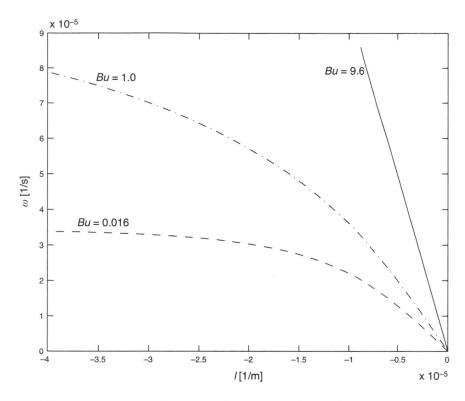

**FIGURE 5.8.** Dispersion curves (coastal-trapped wave frequency $\omega$ as a function of alongshore wavenumber $l$) for three values of the Burger number $Bu$, including the two cases in Figure 5.7. Topography and $f$ are as in Figure 5.7.

These expectations help with interpretation of the computed modal structures for pressure. An idealized pair of calculations uses identical parameters except for the value of the constant $N^2$ (Figure 5.7). With small $N^2$ ($Bu = 0.016$), $\omega = 2.76 \times 10^{-7}$ 1/s, and contours of pressure are virtually vertical over the topography: the wave structure is essentially barotropic. For an identical case except that $N^2$ is much larger ($Bu = 9.6$), $\omega = 10.1 \times 10^{-7}$ 1/s, and the zero contour is almost horizontal, much like a flat-bottom internal Kelvin wave in which the zero contours are exactly horizontal (Figure 5.4 vs. Figure 5.7, right panel). Although with $Bu = 9.6$, the pressure contours away from the coast show considerable depth dependence, the pressure is relatively depth-independent for $x < 10$–20 km, essentially because the natural vertical scale (5.6.7) is large relative to the local water depth, rather than because stratification effects are weaker in shallower water. For perspective, in the case of flat-bottom internal Kelvin waves, the modal structure plays out in the vertical, while for barotropic shelf waves the structure is strictly a function of horizontal cross-shelf distance. With stratification, the modal structure is best thought of as a function of distance along the bottom.

In a continuation of the comparison of large and small $Bu$, the dispersion curves for the two cases are shown in Figure 5.8, along with an intermediate case. The wave for large $Bu$ is essentially nondispersive, just as an internal Kelvin wave would be. Like a barotropic shelf wave, the coastal-trapped wave for small $Bu$ is strongly dispersive, with its frequency approaching a constant value for large negative wavenumbers. Huthnance (1978) showed that for this short-wave limit, the frequency approaches the maximum value of $(Nh_x)$ evaluated along the bottom. Thus, for large wavenumbers, the effects of stratification are not entirely negligible, even with very small $Bu$. Further, for short waves, the structures are relatively bottom trapped, a finding consistent with the scaling (5.6.7), except that for waves shorter (in $y$) than the topographic scale $L^x$, the alongshore wavelength becomes a more appropriate, and shorter, horizontal scale. Note, too, that the long wave's frequency (hence phase speed) is substantially larger with stronger stratification: stratification effectively reinforces the topographic restoring mechanism, so the wave moves faster.

To this point all wave types have been treated in a simplified context with no dissipation or ambient flows. When the frequency is low enough, for example, when $\sigma_F/(\omega H_s) = O(1)$ ($H_s$ being a typical depth over the shelf, and $\sigma_F$ a bottom resistance parameter; see chapter 3), coastal-trapped wave modal structures are severely modified; they have weak velocities over the shelf and relatively stronger currents over the slope (Power et al., 1989; Brink, 2006). Further, when wave frequencies become low enough to be comparable to those of baroclinic Rossby waves in the adjoining ocean, the coastal wave mode couples to the Rossby wave, thus radiating energy offshore (e.g., Clarke and Shi, 1991; Samelson, 2017; chapter 10). In addition, if there is a mean alongshore flow $v_0(x, z)$ that has relative vorticity $v_{0x}$ comparable to $f$, then wave modal structures and propagation can be strongly modified (Brink, 2006). A mean flow also creates the possibility for growing instabilities (chapter 9). Strong frictional or mean flow effects, in turn, make the coastal long-wave modal expansion (see section 5.7) dubious. Finally, isobaths in the real ocean are certainly not straight lines. Topographic irregularities can modify wave propagation and lead to scattering among the wave modes (e.g., Wilkin and Chapman, 1990; Johnson, 1991).

When stratification is sufficiently strong, as defined by Huthnance (1978), coastal-trapped wave dispersion curves can reach $|\omega| \geq |f|$, just as a Kelvin wave would. (Some caution is required in treating $p$ solutions near $\omega = f$, however, because of the presence of a spurious solution at that frequency; Dale et al., 2001). The superinertial situation is rather complicated, because the sloping bottom couples wave motions so that Kelvin-like waves continuously lose energy to internal waves (Dale et al., 2001). The rate of radiative energy loss for either Kelvin or edge waves depends sensitively on matching internal wave scale and frequency.

Very often, shelf current or pressure measurements represent a mixture of wind-forced fluctuations and freely propagating waves (see next section). But sometimes (e.g., Figure 5.1),waves are found propagating freely alongshore in the absence of substantial locally wind-driven motions. A classic example is provided in Figure 5.9, which shows wind vectors, barometrically adjusted coastal sea level,[1] and midshelf, mid-depth

---

[1] Barometrically adjusted sea level is simply a matter of computing the pressure actually felt in the upper water column: air pressure $p_{ATM}$ plus surface-height pressure; that is

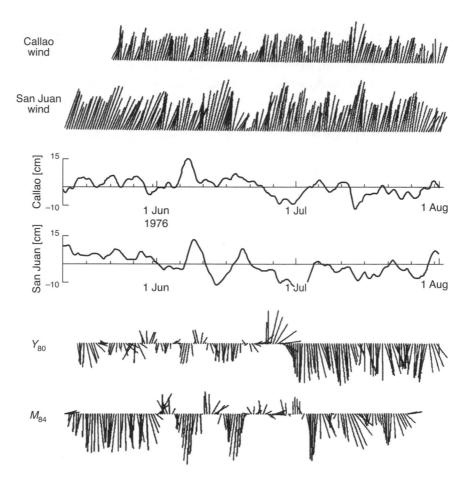

**FIGURE 5.9.** Coastal-trapped alongshore propagation observed along the coast of Peru. Wind records from 12°S (top row) and 15°S (2nd row), adjusted coastal sea level records from near 12°S (3rd row) and 15°S (4th row), and midshelf current measurements from near 12°S (5th row) and 15°S (6th row). Vector records have been rotated so that the local alongshore direction is up/down, and all time series have been smoothed so that sea breeze, tides, and internal waves are removed. For scale, 10 days correspond to 25 cm of sea level, 10 m/s of wind, and 50 cm/s of currents. Adapted from Smith (1978).

currents off the coast of Peru. The first, third, and fifth time series are observations from near 12°S, while the remaining plots represent corresponding observations from near 15°S, about 400 km away. There is no correlation of either wind record (first two subplots) with the ocean pressure or alongshore velocity observations, and cross-shelf currents are

$$\zeta_{adj} = \left[ \frac{p_{ATM}}{(\rho_0 g)} + \zeta \right],$$

where $\rho_0$ is the water density. Using adjusted sea level does *not* involve any assumptions about an "inverse barometer"; that is, there is no assumption that the free surface adjusts to compensate perfectly for air pressure changes.

generally weak. Consistent with geostrophic alongshore flow, local alongshore currents are well correlated with nearby coastal sea level (compare the third with the fifth time series, or the fourth with the sixth). But the important point is that ocean records from 12°S are well correlated with those at 15°S but lead by 2 days. Thus, current and related sea level fluctuations are propagating poleward (with the coast on the left, viewed looking in the direction of propagation—this is the southern hemisphere!) alongshore at a speed of about 200 km/day, in good agreement with the theoretical phase speed for an $n = 1$ mode with local topography and stratification (Brink, 1982). After Smith (1978) detected these waves, they were found to originate as equatorial eastward energy propagation associated with wind-driven Yanai waves (Enfield et al., 1987; see also section 10.2). Once the eastward-moving equatorial disturbance encounters the west coast of South America, it splits and propagates poleward in the form of coastal-trapped waves that behave as internal Kelvin waves and then evolve to be more influenced by topography as they propagate poleward (as $|f|$ increases and $Bu$ decreases).

## 5.7.    The Coastal Long-Wave Approximation and Wind Forcing

The properties of subinertial trapped waves with large alongshore scales (small $l$) hint at some practical simplifications that can occur in this limit. The resulting approximation is rationalized here, following the groundbreaking work of Gill and Schumann (1974) and Gill and Clarke (1974), who showed how useful this concept can be.

First, consider some scales of shelf circulation. One is that the shelf-slope topography (width $L^x$, typically $< 200$ km) is narrow compared with the scales ($L^y$: alongshore) of the weather patterns that often drive motions over the shelf. Thus, $L^x/L^y \ll 1$. Next, typical time scales of wind forcing $\hat{t}$ are a few days or longer, so that often $\hat{t}f \gg 1$ at mid-latitudes. Taken together, these time- and length scale assumptions confine our attention to the area near the origin of the dispersion plot (Figure 5.8) where the curves are very nearly linear (hence, propagation is nondispersive). Next, from the continuity equation (2.4.10c), $u_x = O(v_y)$, so that $\hat{u}/\hat{v} = O(L^x/L^y) \ll 1$, where $\hat{u}, \hat{v}, \hat{p},$ and $\hat{\tau}$ are representative scales of the current components of pressure and of wind stress

These scales can be applied to the depth-averaged linearized equations of motion in the absence of density stratification (i.e., very small $Bu$). The starting point is the shallow-water equations (2.4.10), using pressure $p = \rho_0 g \zeta$, the wind-stress vector ($\tau_0^x$, $\tau_0^y$), and the bottom stress formulation (3.4.8) where $\sigma_F$ is a frictional coefficient (see chapter 3). Thus (dropping the overbars),

$$u_t - fv \quad = -\frac{1}{\rho_0}p_x \quad +\frac{1}{\rho_0 h}\tau_0^x \quad -\sigma_F u/h \qquad (5.7.1a)$$

with magnitudes

$$O[\hat{v} L^x/(\hat{t}L^y)] - O(\hat{v} f) = O[\hat{p}/(\rho_0 L^x)] + O[\hat{\tau}/(\rho_0 H)] - O[\hat{v}\,\sigma_F L^x/(HL^y)] \qquad (5.7.1b)$$

and (for the alongshore equation)

$$v_t + fu \qquad = -\frac{1}{\rho_0}p_y \qquad +\frac{1}{\rho_0 h}\tau_0^y \qquad -\sigma_F v / h \qquad\qquad (5.7.1c)$$

with

$$O(\hat{v}/\hat{t}) + O(\hat{v}\,f\,L^x\,/\,L^y) = O[\hat{p}\,/\,(\rho_0\,L^y)] + O[\hat{\tau}\,/\,(\rho_0 H)] - O(\hat{v}\,\sigma_F\,/\,H). \qquad (5.7.1d)$$

Because typically $L^x/L^y \ll 1$ and $\hat{t}f \gg 1$, the magnitude $L^x/(\hat{t}fL^y) \ll 1$ is doubly small, so the $u_t$ term in (5.7.1a) (slow acceleration of the weak cross-shelf flow) is negligible relative to $fv$. Further, over the midshelf, it is often reasonable to assume that the frictional term is no larger than the time variation term, so that $\sigma_F/(fH) \leq O(1/(\hat{t}f)) \ll 1$, where $H$ is a typical depth.[2] Thus, the bottom friction term in (5.7.1a) is also doubly small, hence negligible.

Conversely, in the alongshore momentum equation (5.7.1.c), the $v_t$, $fu$, and $\sigma_F v/h$ terms are of the same magnitude (because $1/(\hat{t}f)$, $L^x/L^y$, and $\sigma_F/(fH)$ are all small), and all are small relative to $f\hat{v}$ in (5.7.1a). Likewise, $p_x$ is larger than $p_y$ by a factor of $(L^y/L^x)$.

Finally, there are the wind-stress terms, where it is assumed that the alongshore and cross-shelf components are of the same magnitude. If $\hat{\tau}/(\rho_0 H)$ is $O(\hat{v}f)$, then the cross-shelf wind stress would enter (5.7.1a), and the wind stress would provide the only larger term in the alongshore equation (5.7.1d), but this is inconsistent. Instead, it is consistent to say that $\hat{\tau}/(\rho_0 H)$ is $O(\hat{v}fL^x/L^y)$ so that it fits comfortably into (5.7.1c). If this is the case, then the cross-shelf wind-stress term in the cross-shelf equation (5.7.1a) is negligible; only the alongshore wind-stress component in (5.7.1c) plays a role. The unimportance of cross-shelf winds is consistent with the knowledge that these winds drive alongshore net Ekman transport, which, not encountering any barriers, is not very effective at driving interior currents (see chapter 4). Indeed, the settings where cross-shelf winds are known to be important clearly either violate the present assumptions (through short time scales like a hurricane passage or through short alongshore scales, as in a cross-shelf wind jet) or occur because of the breakdown of the simple frictional parameterization (3.4.8) in very shallow water (see section 6.2). It is often assumed that the wind stress is uniform in $x$, with the understanding that its cross-shelf variation is on the same large scale as alongshore variations (although this assumption is not necessary, nor is it always valid in nature).

The net result is that (5.7.1a) reduces simply to a statement that the alongshore flow is in geostrophic balance, so that all the predictive physics is encapsulated in the alongshore equation (5.7.1c). Thus, the depth-averaged *coastal long-wave*[3] equations of motion for the barotropic (constant-density) case become

---

2 This approximation may be reasonable in deeper waters, but it is clearly invalid once the water depth is comparable to, or less than, the thickness of a turbulent boundary layer. See chapter 3.

3 This approximation is consistently called "coastal long wave" here to distinguish it from the long gravity wave assumption (that water is shallow compared to wavelength) that justifies the use of a hydrostatic approximation (see chapter 6). The coastal long-wave approximation assumes hydrostatic conditions from the outset.

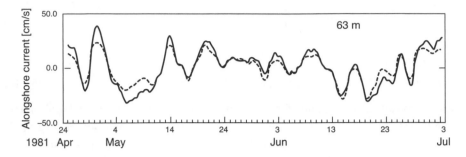

**FIGURE 5.10.** Observations demonstrating geostrophy for alongshore flow over the shelf: alongshore velocity (dashed line) and $(\rho_0 f)^{-1} p_x$ (solid line) at 63 m depth at midshelf off northern California. The time series have been smoothed to remove tides and internal waves, and means have also been removed, because the mean pressure gradient is hard to estimate to the needed accuracy. Adapted from Brown et al. (1987).

$$-fv = -\frac{1}{\rho_0} p_x, \tag{5.7.2a}$$

$$v_t + fu = -\frac{1}{\rho_0} p_x + \frac{\tau_0^y}{\rho_0 h} - \frac{\sigma_F}{h} v, \tag{5.7.2b}$$

$$(hu)_x + (hv)_y + p_t/(\rho_0 g) = 0. \tag{5.7.2c}$$

When density stratification is important, the coastal long-wave approximation once again leads to geostrophic alongshore flow, but the stress terms do not enter in the same way because of the need to deal directly with spatially distinct surface and bottom boundary layers. In any case, the coastal long-wave approximation excludes all near- and super-inertial wave motions (section 5.12), so the mathematics is considerably simplified.

The preceding scaling leads to the conclusion that the alongshore flow is in geostrophic balance. This is indeed a very good approximation over the continental shelf on time scales longer than a few days (e.g., Figure 5.10). If, as is often the case, alongshore flow is in the same direction all across the shelf, then geostrophy is consistent with surface elevation or depression being greatest at the coast and decreasing offshore, as is generally found with the $n = 1$ coastal trapped wave mode. If this is true, time variations in the cross-shelf pressure gradient (hence alongshore velocity on the shelf) are correlated with coastal sea level variability, as is indeed often observed in the ocean (e.g., Figure 5.9).

Now consider the wind-forced problem. The barotropic system (5.7.2), which encapsulates the coastal long-wave approximation, can be reduced to a single equation for pressure:

$$0 = (hp_{xt})_x + fh_x p_y + (\sigma_F p_x)_x - \frac{f^2}{g} p_t - f\tau_{0x}^y, \tag{5.7.3}$$

subject to no normal flow at the coast:

$$0 = p_{xt} + fp_y - \frac{f}{h}\tau_0^y + \frac{\sigma_F}{h} p_x \qquad \text{at } x = 0. \tag{5.7.4}$$

This coastal condition states that the sum of depth-averaged interior transport (first two terms), surface Ekman transport (third term), and bottom Ekman transport (last term) vanishes.

As a first step, consider the free wave solutions to (5.7.3–5.7.4) in the absence of forcing and dissipation $(\sigma_F = \tau_0^y = 0)$, so that

$$0 = (hp_{xt})_x + fh_x p_y - \frac{f^2}{g} p_t, \tag{5.7.5}$$

subject to boundedness for large $x$, and

$$0 = p_{xt} + fp_y \qquad \text{at } x = 0. \tag{5.7.6}$$

This system is separable as

$$p = F_n(x)\varphi_n(y,t), \tag{5.7.7}$$

so that there is an eigenvalue problem for $F_n$ and $c_n$:

$$0 = (hF_{nx})_x + \frac{fh_x}{c_n} F_n - \frac{f^2}{g} F_n, \tag{5.7.8}$$

subject to

$$0 = c_n F_{nx} + fF_n \qquad \text{at } x = 0, \tag{5.7.9a}$$

and (because of boundedness)

$$F_{nx} \to 0 \text{ as } x \to \infty. \tag{5.7.9b}$$

The $(y, t)$ dependence is governed by

$$0 = c_n^{-1}\varphi_{nt} - \varphi_{ny}, \tag{5.7.10}$$

where $\varphi_n$ is thus $\psi(y + c_n t)$, where $\psi$ is any reasonably smooth function. This form illuminates the meaning of the first-order wave equation (5.7.10): it simply states that "what happens here now, happens there later." Note that although this is a wave solution, it is nondispersive, so there is no particular reason for solutions to be periodic, even though dispersive wave problems often give rise to sinusoidal or related forms.

Solutions of the eigenvalue problem (5.7.8–5.7.9) are orthogonal subject to

$$\delta_{nm} = f^{-1}\left[ hF_n F_m \big|_{x=0} + \int_0^\infty h_x F_n F_m dx \right]. \tag{5.7.11}$$

(This condition is found by multiplying eqn. 5.7.8 by $F_m$, subtracting $F_n$ times the $m$ analog of eqn. 5.7.8, integrating over $x$, and applying the boundary conditions.) The seemingly extraneous factor of $f$ assures that wave solutions conserve energy as latitude changes (Brink, 1989).

The forced dissipative problem is approached by multiplying (5.7.8) by $p_t$, subtracting $F_n$ times (5.7.3), integrating over $x$, and applying boundary conditions. A series expansion of the solution

$$p = \sum_{n=0}^{\infty} F_n(x) Y_n(y,t) \tag{5.7.12}$$

is then used in the integral expression, so that applying the orthogonality relation (5.7.11) leads to

$$c_n^{-1} Y_{nt} - Y_{ny} + \sum_0^{\infty} a_{nm} Y_m = b_n \tau_0^y, \tag{5.7.13a}$$

with

$$a_{nm} = \int_0^{\infty} \sigma_F f^{-2} F_{nx} F_{mx} dx, \tag{5.7.13b}$$

and

$$b_n = -F_n(0)/f. \tag{5.7.13c}$$

From this result, other variables can be computed, such as alongshore velocity, using (5.7.2a):

$$v = p_x (\rho_0 f)^{-1} = (\rho_0 f)^{-1} \sum_{n=0}^{\infty} F_{nx}(x) Y_n(y,t). \tag{5.7.14}$$

The infinite sum in (5.7.13a) expresses damping, as well as the fact that bottom friction modifies the inviscid wave modal structure. The derivation of (5.7.13) assumes that $\tau_{0x}^y = 0$. In reality, however, the alongshore wind stress often varies substantially across the shelf (e.g., Beardsley et al., 1987) and inclusion of the resulting wind-stress curl, $\tau_{0x}^y$, modifies the structure of shelf currents (López-Mariscal and Clarke, 1993). With density stratification, $F_n$ becomes a function of both $x$ and $z$, and (5.7.13a) can again be derived, although the orthogonality statement (5.7.11) and the definitions of $a_{nm}$ and $b_n$ are somewhat modified (e.g., Brink, 1989). While the formulation (5.7.13) accounts for a wave mode decaying owing to bottom friction, it is also widely recognized that the waves can lose energy owing to scattering off topographic variations (e.g., Brink, 1980; Wilkin and Chapman, 1990; Kelly and Ogbuka, 2022) or mean current variations (Yankovsky and Chapman, 1996). It is difficult to account for these effects in the context of the present coastal long-wave formulation except under very restrictive conditions (e.g., Brink, 1986).

While (5.7.13) is a remarkably simple result, it is a series solution, so its convergence properties are important. Clarke and Van Gorder (1986) carefully investigated this question and carried out illuminating, illustrative calculations. One important finding was that however quickly the solution for pressure (5.7.13a) might converge, the expressions for other properties, such as alongshore velocity (5.7.14), since they include spatial derivatives, converge more slowly, if at all.[4] Thus, Mitchum and Clarke (1986b) found, for the

---

4 This point can be illustrated by considering a function $\psi(x)$ that is expressed as a convergent series:

$$\psi = \sum_n \Phi_n \sin(nx).$$

west Florida shelf, that using four modes works well for pressure but that at least seven are needed to reach a comparable quality for alongshore velocity. Further, it is not necessary that the lowest modes have the largest amplitude (e.g., Illig et al., 2018), although that often appears to be the case. That convergence is an issue under realistic conditions is perhaps not surprising. With no friction (so that the modal coupling in eqn. 5.7.13a vanishes), this is already a concern. Including friction, hence the coupling, presents the challenge of expressing a frictional solution in terms of modes generated as part of a frictionless theory. Given all these considerations, it is impressive that some remarkably good results have sometimes been found using a small number of modes. Certainly, caution is always warranted in assessing whether enough modes are included in a calculation.

Very often, the off-diagonal terms in (5.7.13a) are ignored, although this is hard to justify theoretically (Clarke and Van Gorder, 1986). This leaves the damped forced equation

$$c_n^{-1}Y_{nt} - Y_{ny} + a_{nn}Y_n = b_n\tau_0^y,$$
(5.7.15)

which is readily solved using the method of characteristics by defining

$$\xi = y + c_n t,$$
(5.7.16a)

$$\eta = t,$$
(5.7.16b)

so that, by changing to the new variables, (5.7.15) becomes

$$Y_{n\eta} + c_n a_{nn} Y_n = c_n b_n \tau_0^y,$$
(5.7.16c)

and thus

$$Y_n = \int_{-\infty}^{t} e^{-a_{nn}c_n\eta} c_n b_n \tau_0^y d\eta \big|_{(y+c_n t) = \text{constant}}.$$
(5.7.16d)

It is important to remember that the integral in (5.7.16d) is not evaluated at a fixed $y$ location but that it involves $y$ values along characteristics, that is, along lines of constant $(y + c_n t)$. An example of using this technique is presented in section 5.8. But what is remarkable here is that the wind-forced response, a function of $x$, $y$, and $t$, can be found in terms of straightforward calculations of modal structures and the simple solution (5.7.16d).

The form of (5.7.13a) helps illustrate an important point in simple terms. Imagine that there is no bottom friction, so that $a_{nm} = 0$, and that

---

Then, the derivative is

$$\psi_x = \sum_n n\Phi_n \cos(nx);$$

that is, the coefficients of the series for $\psi_x$, $n\Phi_n$, include a factor that increases with $n$. Thus, the expansion of $\psi_x$ is expected to converge more slowly than that for $\psi$, if at all.

$$\tau_0^y = \tau_A \sin(\omega t) \qquad \text{for } y < 0, \tag{5.7.17a}$$

$$\tau_0^y = 0 \qquad \text{for } y > 0. \tag{5.7.17b}$$

The particular, or forced, solution for $y < 0$ is

$$Y_{np} = -\frac{b_n c_n \tau_A}{\omega} \cos(\omega t), \tag{5.7.18}$$

and $Y_{np} = 0$ for $y > 0$. The homogeneous, or free wave, solution is

$$Y_{nh} = \psi(y + c_n t) = \frac{b_n c_n \tau_A}{\omega} \cos\left[\frac{\omega}{c_n}(y + c_n t)\right] \qquad \text{for } y < 0, \tag{5.7.19a}$$

$$Y_{nh} = 0 \qquad \text{for } y > 0, \tag{5.7.19b}$$

because the waves in the northern hemisphere propagate only toward $-y$. The form and amplitude in (5.7.19a) are chosen so that $Y_n = Y_{np} + Y_{nh} = 0$ at $y = 0$. The total solution (forced plus free) for $y < 0$ can thus be written as

$$Y_n = -\frac{2b_n c_n \tau_A}{\omega} \sin\left(\frac{\omega y}{c_n}\right) \sin\left[\omega\left(2t + \frac{y}{c_n}\right)\right]. \tag{5.7.20}$$

The important point here is that the alongshore propagation (as opposed to amplitude modulation) is described by the second sine function in (5.7.20): it shows alongshore propagation of the net signal at a speed of $2c_n$! This net speed reflects the sum of equal-amplitude waves that propagate with infinite speed (the forced "wave") and with speed $c_n$ (the free wave). Although this example is highly idealized, it makes the point that the total signal that one observes in nature is usually a comparable mixture of free waves (that propagate at speed $c_n$) and forced waves that may or may not show alongshore propagation, at a speed determined by the local meteorology. The combined waves represent a signal that does not propagate with the free wave speed. Only in unusual circumstances, such as those described in section 5.1 or in Figure 5.9, is the wind forcing confined enough in space and time that away from the forcing region, ocean observations reveal propagation at only the free wave speed.

As an aside, it is helpful to revisit (5.7.8) using the step topography (5.5.4) in the coastal long-wave, rigid-lid, barotropic limit. In this case, there is only one shelf wave mode because the bottom is discontinuous, but this solution is readily found. Specifically, the phase speed for the long shelf wave is

$$c_1 = \frac{fL(H_0 - H_1)}{H_0}. \tag{5.7.21}$$

This result, although greatly simplified, provides a handy way to estimate the magnitude of the phase speed of the first-mode barotropic shelf wave. In many cases, the shelf is far shallower than the deep ocean ($H_1 \ll H_0$), so the solution simplifies even further to $c_1 \approx fL$; that is, barotropic shelf waves travel faster at higher latitudes and when the shelf is wider.

## 5.8.    Solving for Large-Scale, Wind-Forced Motions

It is useful to consider an illustrative wind-forced problem and to solve it using the method of characteristics. The example here closely follows the very lucid presentation of Allen (1976). Carton (1984) went beyond the present discussion to explore different forcing configurations and to elucidate the consequences of accounting for multiple wave modes.

To proceed, say that the alongshore wind stress is turned on suddenly and is confined to an alongshore strip of length $L$:

$$\tau_0^y = \tau_A \qquad \text{for } t > 0, \ -L < y < 0, \tag{5.8.1a}$$

$$\tau_0^y = 0 \qquad \text{otherwise}, \tag{5.8.1b}$$

and $\tau_A$ is a constant. For simplicity, frictional damping is ignored, so, for each mode, (5.7.16c) becomes

$$Y_{n\eta} = c_n b_n \tau_0^y \qquad \text{for } y + c_n t = \text{constant}, \tag{5.8.2}$$

and $\eta$ is time along a characteristic. To make this tangible (see eqn. 5.7.12), $F_n Y_n$ is the mode $n$ contribution to pressure, so for convenience, one might think of $Y_n$ as being a measure of the mode's contribution to, say, pressure at midshelf.

To visualize the problem, it is useful to sketch it out in the $(y, t)$ plane (Figure 5.11A). The region in this space where there is wind forcing lies in a band between $y = -L$ and $y = 0$. Superimposed on this plot is a sampling of characteristics, that is, lines where $y + c_n t$ is a constant and along which (5.8.2) is to be integrated. Arrows along a couple of the characteristics indicate the sense in which (5.8.2) is integrated. If a characteristic (e.g., $y + c_n t = -3L/2$) does not pass through the range where $\tau_0^y \neq 0$, then no forcing is experienced, and $Y_n = 0$ everywhere along that characteristic. Likewise, $Y_n = 0$ for $t < 0$, because there is no forcing anywhere then. Finally, $Y_n = 0$ for $y > 0$ and $t > 0$. This follows because information in the northern hemisphere moves only toward $-y$, so the ocean here cannot "know" that the forcing exists. These regions where $Y_n = 0$ are denoted by darker shading around the edges in Figure 5.11A.

Actual solutions are now obtained by time-integrating along the characteristics. For example, say that $y + c_n t = 0$. This is a special choice, because this curve passes through the $y = 0$ edge of the forcing at $t = 0$. Integrating along this curve,

$$Y_n = c_n b_b \tau_A t = -b_n \tau_A y \qquad \text{for } 0 < t < L/c_n, \tag{5.8.3a}$$

and $Y_{n\,\eta} = 0$ for $t > L/c_n$ (or, equivalently, for $y < -L$), so that

$$Y_n = c_n b_n \tau_A \, L/c_n = b_n \tau_A \, L \qquad \text{for } t > L/c_n. \tag{5.8.3b}$$

A similar result is obtained for all characteristics passing through $y > 0$ when $t \geq 0$. For example, consider the case of $y + c_n t = L/2$. This characteristic does not encounter any forcing until $t = L/(2c_n)$, so that

$$Y_n = 0 \qquad \text{for } t < L/(2c_n), \tag{5.8.4a}$$

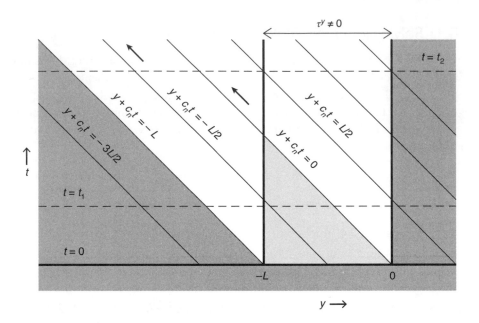

**FIGURE 5.11A.** Schematic illustrating first-order wave equation characteristics (the diagonal lines). Several of the characteristics are labeled by their defining equations, and arrows indicate the sense of integration along the characteristic. Dark shading around the edges represents regions where $Y_n = 0$. The triangular lightly shaded region in $-L \leq y \leq 0$ is where the solution is temporarily independent of $y$.

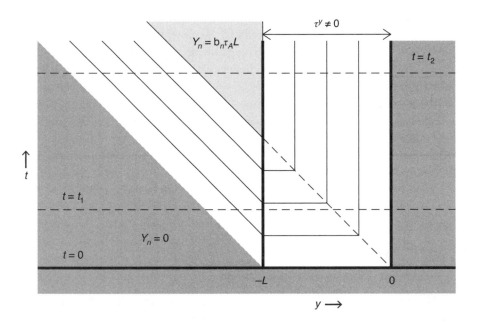

**FIGURE 5.11B.** Contour plot of the $Y_n$ solution that accompanies the characteristic diagram of part A. The darkly shaded region has $Y_n = 0$, while the lightly shaded region is a plateau where $Y_n$ reaches its maximum magnitude, $Y_n = b_n \tau_A L$. The diagonal dashed line is the characteristic $y + c_n t = 0$.

$$Y_n = c_n b_n \tau_A [t - L/(2c_n)] = -b_n \tau_A y \qquad \text{for } L/(2c_n) < t < 3L/(2c_n), \qquad (5.8.4.b)$$

$$Y_n = c_n b_n \tau_A L/c_n = b_n \tau_A L \qquad \text{for } t > 3L/(2c_n). \qquad (5.8.4c)$$

The result (5.8.4b) is expressed by the $Y_n$ contours parallel to the time axis in Figure 5.11B.

Choosing a different characteristic that passes through, say, $y = -L/2$ at $t = 0$ yields a somewhat different result. In this case, the characteristic passes out of the forcing region before $t = L/c_n$, so that

$$Y_n = 0 \qquad \text{for } t < 0, \qquad (5.8.5a)$$

$$Y_n = c_n b_n \tau_A t \qquad \text{for } t < L/(2c_n), \qquad (5.8.5b)$$

and

$$Y_n = b_n \tau_A L/2 \qquad \text{for } t > L/(2c_n). \qquad (5.8.5c)$$

This particular characteristic is representative of the range $-L < y < -c_n t$ (the triangular light-shaded region of Figure 5.11A) in that the response at a given time is the same along each characteristic as long as $y > -L$. This result is expressed by the $Y_n$ contours parallel to the $y$ axis in Figure 5.11B. Physically, for this $y$ range, the ocean responds initially as if the forcing extends over all $y$. Only once a wave can propagate in from the $y = 0$ boundary does the ocean "know" that the forcing region has finite $y$ extent.[5] After that time, the solution $Y_n$ has an alongshore gradient (e.g., 5.8.4b). For $y < -L$, where there is no wind forcing, $Y_n$ is constant along a characteristic at whatever value it has at $y = -L$.

The individual solutions along characteristics can now be sorted out into plots of $Y_n$ as a function of $y$ at different times (Figure 5.12). These are effectively slices through Figure 5.11B. At time $t = t_1 < L/c_n$ (the lower dashed horizontal line in Figures 5.11A and B), there has not been enough time for a wave originating at $y = 0$ to cross the entire forcing domain. Thus, the range with $-L < y < -c_n t_1$ still behaves as if the forcing were completely uniform alongshore. The maximum $Y_n$ for any $t < L/c_n$ is less than the maximum (5.8.3b or 5.8.4c), which occurs only once characteristics pass through the entire $y$ forcing range of 0 to $-L$. On the $-y$ side of the forcing region, characteristics originating at $t = 0$ at $y = -L$ to 0 define the boundaries of a ramp downward to zero for $y < -L$:

$$Y_n = b_n \tau_A (c_n t + L + y) \qquad \text{for } -c_n t - L < y < -c_n t. \qquad (5.8.6a)$$

At later times (such as $t_2$ in Figures 5.11A and B), the plateau $Y_n$ is simply the constant value that matches the $y > -L$ solution:

$$Y_n = b_n \tau_A L \qquad \text{for } -c_n t < y < -L. \qquad (5.8.6b)$$

---

5 This provides an important insight on the validity of assuming two-dimensionality in time-dependent models. For forcing that is uniform over a limited $y$ scale, response can appear two-dimensional only over relatively short times. See also chapter 4.

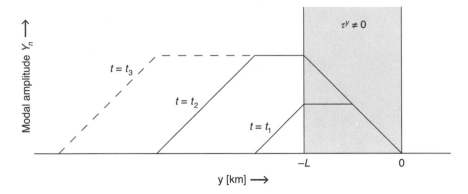

**FIGURE 5.12.** Plots of modal amplitude $Y_n$ as a function of $y$ at three different times. Times $t_1$ and $t_2$ are denoted in Figures 5.11, while $t_3$ is a yet larger time beyond the range of Figures 5.11. The shaded area represents the $y$ range over which wind forcing occurs.

Note that the $y < -L$ range where $Y_n$ is a constant keeps growing with time, as shown by the $t_2$ and $t_3$ lines in Figure 5.12.

It is now possible to interpret the results for, say $t = t_2$ in Figure 5.12. Taking $Y_n$ to be like pressure or alongshore velocity, there is now a uniform alongshore pressure gradient in the range $-L < y < 0$. Physically, the cross-shelf wind-driven Ekman transport in the coastal ocean is compensated for by an alongshore current that strengthens toward $-y$ and by a geostrophic cross-shelf flow associated with the alongshore pressure gradient. Beyond the forcing region, for $y < -L$, the current extends alongshore unchanged, $Y_{ny} = 0$, until $y < -c_n t$, when it begins to weaken, so that $Y_{ny} \neq 0$ in this "nose" region. This nose has an alongshore pressure gradient, hence cross-shelf flow, that drains the diminishing alongshore current offshore. For $y < -L - c_n t$, $Y_n = 0$, because there has not been enough time for information about the wind forcing to reach this far.

An interesting aspect of this solution is that although the forcing continues forever and there is no friction, the response amplitude does not increase indefinitely. Rather, with time, energy is radiated away from the forcing region, as expressed by the continually expanding alongshore range over which the solution extends (Figure 5.12).

In conclusion, the method of characteristics allows remarkably simple solutions to this problem. In a more realistic case with dissipation, there would be a weaker response and an exponential decay of the solution outside the forcing region, but the bounding roles of the various characteristics would remain the same. One complexity arises in mapping the results back into $(y, t)$ space to obtain results in a more familiar form, like Figure 5.12. Further, the bookkeeping becomes somewhat complicated if the frictional couplings in (5.7.13a) are included (Clarke and Van Gorder, 1986). Nevertheless, the underlying simplicity remains: information is accumulated and transferred only in the direction of long coastal-trapped wave propagation, so that many properties of the solution can be seen simply from considering diagrams such as Figure 5.11A.

## 5.9.    Some Coastal-Trapped Wave Applications

### Hindcasts

A number of authors use long coastal-trapped wave theory to hindcast continental shelf currents and/or pressure. Some of these calculations represent impressive, comprehensive efforts to arrive at optimal solutions, usually based on the stratified equivalent of (5.7.13) (e.g., Mitchum and Clarke, 1986b; Chapman, 1987; Illig et al., 2018). Authors have explored the various choices made in performing these wave hindcasts and have often found it necessary to use more than two modes to optimize the solutions. Almost invariably, the hindcasts provide time series that are very well correlated with observed pressure and alongshore currents, but usually the models somewhat underpredict the amplitudes of these observed time series (e.g., Figure 5.13). The models almost invariably do a terrible job of hindcasting temperature and cross-shelf currents, yielding uncorrelated time series with amplitudes far too weak relative to observations. That the hindcast $u$ amplitude is small is built into the coastal long-wave theory, of course. Other reasons for this failing include the assumed thin turbulent boundary layers, the lack of horizontal advection (very likely important for temperature especially), linearity, and the assumption of hydrodynamic stability (see chapter 9). So, it often appears that forcing by large-scale alongshore winds, by itself, is not an adequate explanation for observed $u$ or $T$ (hence $\rho$) variability.

One measure of how well the wave models perform involves creating a strictly empirical, statistically optimized hindcast based on regressions with wind and other inputs of

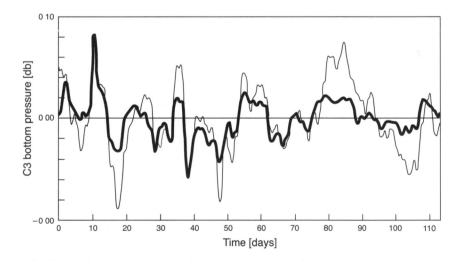

**FIGURE 5.13.** Results from a coastal-trapped wave hindcast. Observed (light line) and hindcast (heavy line) midshelf bottom pressure off northern California. The hindcast was made using linear coastal-trapped wave theory and observed wind stresses. Means have been removed from both time series, and smoothing has removed tides and other higher-frequency effects. Adapted from Chapman (1987).

the sort that might go into a wave calculation. In principle, this is a standard that is not expected to be matched by theory (which is not statistically optimal), given the same sorts of inputs. Chapman et al. (1988) computed such an optimized empirical time series and then compared it with results of coastal long-wave hindcasts off California. They found that, within error, the wave model performed as well as the optimal empirical model and was better in the sense that it did so well without any sort of data-based tuning. This was indeed a high level of performance, especially for such a simple model.

## Stochastic Models

The solutions to the forced damped first-order wave equation (5.7.15) can be used, along with information about the space- and time-lagged correlation function of the wind stress, to predict the space-time correlation function of $Y_n$ (which is proportional to both $v$ and $p$ for a given mode), and the space- and time-lagged correlation between wind stress and $Y_n$ (Allen and Denbo, 1984). They found that the maximum wind-$Y_n$ correlation occurs with winds earlier in time and at a larger $y$ (in the northern hemisphere coordinate system, where $x$ increases offshore) than the ocean measurement, a result often found In observations (e.g., Harden et al., 2014). Physically this makes sense, because this is the direction from which wave signals propagate. The magnitude of the correlation, and the space-time lags to maximum correlation, depend critically on the frictional damping, in the sense that larger friction results in a stronger correlation, because less incoherent energy is propagating in from far away. The strength of the response depends on the dominant direction of weather system propagation: if weather systems generally translate in the same sense as free wave propagation (as they often do off the U.S. west coast), the response is relatively resonant and stronger than if weather systems move in the opposite sense (as they do off the U.S. east coast south of New England). Stating it another way, coastal long-wave resonances depend on matching propagation speeds rather than frequencies or wavenumbers. It has since been shown (Chapman et al., 1988) that the stochastic model more consistently reconciles observations with input and theoretical parameters if three wave modes, rather than just one, are used.

These results are consistent with correlation functions estimated from actual sea level and wind data from off the west coast of North America (Figure 5.14). The broad features of these plots are identical to findings of the stochastic theory. To begin, the wind-stress autocorrelation $\hat{R}_{\tau\tau}$ diagrams show a weak tendency for northward weather system propagation north of 41.8°N but no preferred direction south of that latitude. Peak correlations between winds and sea level $\hat{R}_{\tau\zeta}$ occur with wind about 300–500 km to the south of the ocean measurement and 1–2 days earlier. The tilt of the correlation diagrams for sea level records with each other, $\hat{R}_{\zeta\zeta}$, indicates that the sea level fluctuations propagate northward at a speed that reflects a mixture of free and forced waves (as pointed out in section 5.7). One really useful aspect of these results is the very clear demonstration of the dependence of correlation patterns of observed quantities on the wave underpinnings.

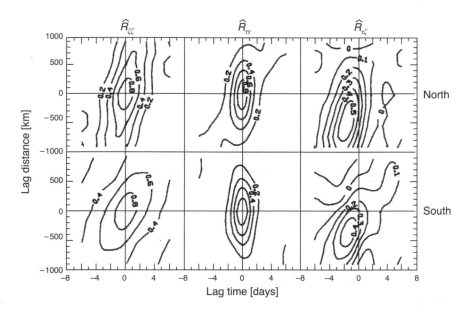

**FIGURE 5.14.** Space- and time-lagged correlation functions computed from wind stress and coastal sea level observations off the west coast of North America (summer 1973). These were computed separately for northern (upper row) and southern (lower row) zones. The dividing line between north and south is 41.8° N. Alongshore spatial lags are on the vertical axis, and time lags on the horizontal. The sign convention is such that an upper right/lower left tilt represents northward propagation. The left-side plots are correlations among sea level records, the central plots are correlations among alongshore wind stress records, and the right-side plots are correlations between alongshore wind stress and sea level. In this coordinate system, free coastal-trapped waves propagate toward positive *y*. Adapted from Allen and Denbo (1984).

## Tides on the Shelf

Current measurements on the continental shelf immediately north of the Strait of Juan de Fuca (off western North America around 49°N) show that diurnal tidal currents are anomalously strong there compared with south of this major disruption in the coastline. Crawford and Thomson (1984) examined these currents in light of the dispersive coastal-trapped wave modes that exist at the subinertial $K_1$ and $O_1$ (see section 7.2) tidal frequencies. They found that a combination of the $n = 0$ barotropic Kelvin-like mode and the $n = 1$ coastal-trapped mode provided an excellent fit to their measurements. The $n = 0$ mode, with its large horizontal scale, hence weak pressure gradient, dominates the pressure (including sea level) variance, whereas it has very weak associated currents. The $n = 1$ mode has shorter horizontal scales (roughly the shelf width) and thus has strong pressure gradients and dominates the current variability. Evidently, tidal flow disruptions associated with the strait excite the $n = 1$ mode, which then propagates northward and favors the existence of strong diurnal shelf currents.

## 5.10.   Circular Geometries

Subinertial trapped waves can also be found in a variety of circular geometries. For example, in the case of an idealized island (Brink, 1999), the cylindrical coordinate $(r, \varphi)$ equivalents of (5.6.1 and 5.6.2) apply, where $r$ is the radial coordinate and $\varphi$ the azimuth. The new twist in this case is the requirement that in addition to the familiar $(r, z)$ discretization, waves are also quantized in the azimuthal direction, so the wave frequency is also quantized. Solutions take the form

$$p = F_{nm}(r, z) \sin(m\phi - \omega_{nm}t), \tag{5.10.1}$$

where $(n, m)$ are both integers. That is, the "alongshore" wavenumber $m$ is discrete, as opposed to the situation with a straight coast, where alongshore wavenumber $l$ can vary continuously. Crudely stated, an integral number of wavelengths must fit around the island. For some $(n, m)$ pairs, solutions may not exist if the island is too small, particularly for rapidly propagating lower radial-vertical mode numbers $n$. If depth increases monotonically outward, then $(mf/\omega) < 0$: the waves propagate clockwise (viewed from above) around the island in the northern hemisphere. These modes can be excited by impinging ambient currents, such as tides, or by winds, although direct meteorological forcing appears to be inefficient.

The straight-coast limit (5.6.1) is reached when the island's radius is large relative to the shelf-slope width or to the internal Rossby radius scale. Further, frictional effects need to be strong enough that a wave dissipates before propagating around the island and returning to its point of origin. There is no coastal long-wave limit until the island is large enough that the straight-coast approximation applies. It is worth noting that Earth's largest islands, such as Australia or Greenland, have sufficiently irregular topography (especially if one follows a deeper isobath, such as 1000 m) that it is difficult to use a circular approximation effectively. The circular idealization appears to apply best to smaller, isolated islands, such as Bermuda where Hogg (1980) detected discrete trapped-wave resonances.

Other circular geometries include idealized seamounts (that have no surface expression) and lakes (that are enclosed). Both these cases have nonzero depth at the center of the coordinate system. This simply requires the replacement of a coastal boundary condition like (5.6.2) with $p = 0$ at $r = 0$. Again, these modes are doubly quantized and propagate in the same sense expected for long coastal-trapped wave modes. Software for computing trapped waves in circular geometries under general circumstances can be found for islands at https://darchive.mblwhoilibrary.org/handle/1912/10526 and for seamounts or basins at https://darchive.mblwhoilibrary.org/handle/1912/10528.

## 5.11.   The Steady Barotropic Limit

In the coastal long-wave limit, one expects information about forcing (by the alongshore wind stress, for example) to propagate alongshore in only one direction. Thus, once a frictional steady state emerges, it, too, will spread only "down-wave" from where forcing is applied; that is, in the northern hemisphere, forcing at $y=0$ will be felt only for $y<0$. This is indeed the case in models of the mean flow. Consider the steady-state limit of the barotropic coastal long-wave governing equations (5.7.3–5.7.4) and assume that the wind stress does not vary across the shelf:

$$0 = (\sigma_F p_x)_x + f h_x p_y, \tag{5.11.1}$$

with

$$0 = p_y - \frac{1}{h}\tau_0^y + \frac{\sigma_F}{fh}p_x \qquad \text{at } x = 0. \tag{5.11.2}$$

The physical balance in (5.11.1) juxtaposes the tendency to create vorticity via vortex stretching (associated with cross-isobath motions: the $h_x$ term) with the frictional dissipation of the resulting relative vorticity. Alternatively, one could think of (5.11.1) in terms of vertical velocity near the bottom: barotropic cross-isobath flow demands a vertical motion, and this is balanced by Ekman pumping associated with the curl of the bottom stress.

As Csanady (1978) pointed out, there is a very useful analogy here to heat conduction. This follows if $\sigma_F$ and $h_x$ are constant, and $q = -y$ is defined. Then, (5.11.1) becomes the heat equation

$$p_q = K_A p_{xx}, \tag{5.11.3}$$

with

$$K_A = \frac{\sigma_F}{fh_x}. \tag{5.11.4}$$

Thus, the negative alongshore direction plays the role of time, and the cross-shelf direction is like distance along a conducting bar. The "diffusivity" $K_A$ increases with bottom friction and decreases with bottom slope and $f$. This analogy allows the use of known results and intuition about heat conduction. For example, if a uniform alongshore wind stress is applied everywhere that $y<0$, then the coastline acts like the end of a bar where a constant (in "time") heat flux is applied. The farther one travels toward $-y$ (larger "time"), the more spread out the pressure field ("heat") becomes. The rate of cross-isobath spreading is governed by the diffusivity (5.11.4): it increases with the effects that allow easier cross-shelf flow in a rotating system: friction, slower rotation, or a flatter bottom.

Csanady called this steady coastal long-wave limit "the arrested topographic wave," an apt name that relates the transient shelf waves to the steady flow they establish. One can, however, ask about the various assumptions involved here. If the coastal long-wave approximation is not made, there is an additional term in (5.11.1), $(\sigma_F p_y)_y$, that allows the solution to penetrate in the positive $y$ direction, but it does not spread far because energy propagation in that direction is slow and readily damped, related as it is to the short shelf waves which propagate energy slowly in the sense opposite to the phase speed. A version of this steady theory has also been derived for highly diffusive cases that include density stratification (McCreary and Chao, 1985). While stratification modifies the barotropic results, it does not seem to cause any new phenomenology for the linear problem. So, all considered, this arrested limit is a very useful idealization.

## 5.12.   Superinertial Motions: What the Coastal Long-Wave Approximation Misses

The coastal long-wave approximation focuses on changes with time scales longer than a few days. It is worthwhile to consider what is then left out. As a single example, consider the response of an alongshore-uniform, frictionless two-layer (depths $H_1$ and $H_2$), semi-infinite constant-depth ocean responding to a steady alongshore wind turned on suddenly at time $t = 0$. On coastal long-wave time scales of days or longer, a steady cross-shelf circulation develops with (say) offshore Ekman transport in the upper layer far from shore, onshore transport below, and a nearshore region where upwelling occurs, completing the flow pattern (see section 4.5). The nearshore region has the cross-shelf scale of an internal Rossby radius, $c_1/f$, where $c_1^2 = g'H_1H_2/(H_1 + H_2)$ and $g' = g\Delta\rho/\rho_0$. As time goes by, and the interface continues to rise, a baroclinic geostrophic alongshore flow continues to accelerate.

Much is left out by this description (Pettigrew, 1980). Specifically, there are three additional, higher-frequency components of the solution. One, confined to the upper layer, is a spatially uniform, strictly inertial-period oscillation associated with satisfying the initial condition of rest in an infinite unbounded ocean (This is exactly what would occur in a one-dimensional problem; see section 3.5). Next, there are two propagating components, a barotropic Poincaré wave and an interfacial Poincaré wave. Both wavelike solutions are required to assure there is no flow through the vertical coastal wall in either layer. For example, the barotropic mode alone could bring the upper layer to rest at the wall, but then the lower layer would not obey its boundary condition. The excited dispersive waves have frequencies somewhat greater than the inertial to allow for offshore propagation (recall the form 5.2.4). The barotropic mode first propagates very rapidly offshore [at speed comparable to $(g(H_1 + H_2))^{1/2}$], so that once the barotropic wave front has passed, the strictly inertial near-surface component and the barotropic wave component coexist. This combination leads to a roughly 180° phase offset between the upper layer (with its strictly inertial fluctuations) and the depth-independent wave part; that is, the near-surface oscillation is somewhat weakened, and the rest of the water column begins oscillating in a way that maintains a shear at the base of the upper layer. Meanwhile, the baroclinic wave front propagates more slowly offshore from the coast at the internal wave speed near $c_1$. Once it arrives at a given offshore location, it cancels out the near-inertial oscillations in both layers. Thus, at a given offshore location, strongly sheared near-inertial oscillations occur throughout the water column (albeit with an abrupt phase change below the mixed layer) until the baroclinic mode arrives from the coast and (after higher-frequency transients radiate away) terminates the near-inertial fluctuations throughout the water column. After this, the solution is essentially the same as that in section 4.5.

If the same problem is expressed in terms of a continuous density stratification (Kundu et al., 1983), an infinite number of baroclinic modes is excited, but the process is similar, if more gradual. The need to have no normal flow at the coastal boundary means that each internal wave mode is excited to some extent. Careful inspection of these solutions again leads to wave fronts propagating offshore, partially cancelling deeper motions after an initial period of downward, offshore energy propagation.

While near-inertial responses due to winds or other processes might readily be smoothed out of ocean-data time series, the wave motions here are often associated with very substantial velocity shears. These shears, in turn, if large relative to the density stratification in the sense of a small Richardson number, give rise to turbulent mixing. Recent estimates, in fact, suggest that coastally wind-generated internal waves make a nonnegligible contribution to the total global ocean mixing (Kelly, 2019).

## 5.13.   Conclusion

There is a well-established literature concerning wave motions in the coastal ocean. A particularly successful part of this corpus involves long coastal-trapped waves, which can be used remarkably effectively to hindcast and understand sea level and alongshore currents over the shelf. These waves serve as a strong reminder of the inherent asymmetry in the coastal ocean that favors the subinertial propagation of energy in only one direction (i.e., propagating with the coast on the right-hand side in the northern hemisphere). Closely related, azimuthally quantized, waves are also possible around lakes, isolated islands, and seamounts. The trapped-wave propagation spreads the response alongshore, so pressure and alongshore currents over the shelf are not generally driven by local winds, but rather, they reflect forcing earlier in time and distributed over the alongshore direction. While these insights are powerful, important gaps in our knowledge remain. For example, there is presently a good deal of active research concerning trapped, or nearly trapped, waves at superinertial frequencies and their relation to tidal forcing and to turbulent mixing.

# 6

# The Inner Shelf

## 6.1. Introduction: Where Surface and Bottom Conditions Interact

A common definition of the inner shelf is that area where the water is shallow enough that the surface and bottom turbulent boundary layers merge. An alternative definition (Becherer, et al., 2021) is that the inner shelf is where incoming tidal internal waves break, resulting in a high degree of turbulence ("the surf zone for the internal tide"). The two definitions emphasize different drivers (winds vs. internal tides), but in either case, there is no longer an inviscid interior region where turbulent processes play a lesser role. On the inshore side, the inner shelf abuts the surf zone where incoming surface gravity waves break and where wave-driven processes play a dominant role. Either dynamically based definition means that the inner shelf's boundaries move about from day to day and from season to season as stratification, internal tides, surface waves, and currents change. This follows because turbulence generation is dependent on flow properties, stratification, and surface warming (chapter 3). Typically, however, for summertime at midlatitudes, the inner shelf might lie between the 3 and 30 m isobaths.

The inner shelf is important for a variety of reasons, the most obvious being that its position implies a controlling role in the movement of sediments and dissolved materials between land and the broader ocean. As will be seen, the dominance of turbulent processes creates a setting where vertical motions, such as coastal upwelling, can be concentrated. The importance of mixing throughout the water column means that stratification is expected to be weaker on the inner shelf relative to offshore (Figure 6.1). Buoyancy currents (section 8.4) are rooted in the inner shelf and, depending on stratification and transport, may lie entirely within this active domain. Further, processes such as tidal rectification (section 7.4), which are strongest in shallow water, can also dominate in this region.

Thus, there are many reasons, scientifically and practically, to study the inner shelf, yet historically it has been relatively little studied until recent decades. The reasons are not hard to find. One is simply that given the importance of variable and nonlinear

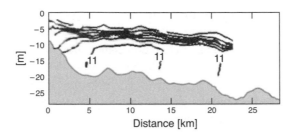

**FIGURE 6.1.** The observed transition from the inner- to midshelf. A cross-shelf section of temperature from July 1998 off the coast of the US state of New Jersey. The contour interval is 2 °C. Note that the temperature stratification is more diffuse in water shallower than 15–20 m. Adapted from Chant et al. (2004).

turbulent processes, it can be conceptually complicated to treat. But the dominant reason for the previous neglect is that this is a particularly difficult region to observe. Operators of seagoing ships are understandably reluctant to enter such shallow water that is often uncomfortably close to land. At the same time, the waters are too far offshore to be accessible by the ingenious, land-based observational approaches (e.g., Thornton and Guza, 1986) used by the surf zone community. Given the shoaling, large-amplitude surface waves, moorings are subject to severe stresses and can fail without extreme precautions. All considered, the inner shelf well earned its nickname as "the bad zone"; however, more sophisticated observational approaches (e.g., Chant et al., 2004; Fewings et al., 2008) have now made the inner shelf accessible and opened the doors to new levels of understanding.

While overlapping surface and bottom turbulent boundary layers are the defining characteristic of the inner shelf, they are not the only unique property. Another is its role as a dynamical transition zone. In the shallow surf zone, wave driving is overwhelmingly dominant. Offshore of the surf zone, and outward across the inner shelf, surface waves gradually become less important as drivers of currents, even as they can remain a factor for upper-ocean processes.

These two defining qualities (full-water column boundary layers and wave forcing) structure the following treatment. At first, waves will be ignored, and the implications of merged boundary layers are explored. Then, surface waves (nonhydrostatic gravity waves) are reviewed before addressing the offshore zonation of different driving agencies and wave forcing of transports. The surf zone in particular is not treated here at a level of detail consistent with the amount of interesting work in that domain, but good introductions can be found elsewhere (e.g., Dean and Dalrymple, 1992; Komar, 1998).

## 6.2.    Boundary Layers in Shallow Water

To begin, boundary layer structure is considered, with emphasis on the transition between deep water (surface and bottom boundary layers separated by a non-turbulent interior region), and the shallow, fully turbulent zone. For simplicity, it is at first assumed that the entire water column is equally turbulent, i.e., that there is a constant eddy viscosity $A$. More realistic turbulent parameterizations would take into account spatial variations, density stratification and variable flow conditions: effects most readily treated in numerical models (below). However, a constant eddy viscosity does allow convenient solutions that give some insight regarding boundary layers in shallow water. Ignoring stratification, momentum advection and alongshore variability for now, the two-dimensional $(x, z)$ equations of motion for a homogeneous ocean are

$$u_t - fv = -\frac{1}{\rho_0} p_x + (Au_z)_z, \tag{6.2.1a}$$

$$v_t + fu = +(Av_z)_z, \tag{6.2.1b}$$

$$u_x + w_z = 0. \tag{6.2.1c}$$

where $(u, v, w)$ are the velocity components in the offshore, alongshore and vertical directions, $(x, y, z)$ are the corresponding coordinates, $t$ is time, $p$ is pressure, $\rho_0$ is a constant density, $f$ is the Coriolis parameter, and subscripted independent variables represent partial differentiation. To further simplify, the flow is considered steady, so that the time-derivative terms are dropped.

Now, the velocity is broken into a boundary layer portion, e.g., $u_E$, and an inviscid portion $u_I$ so that, for example,

$$u = u_I + u_E, \tag{6.2.2}$$

where

$$-fv_I = -\frac{1}{\rho_0} p_x = -g\zeta_x, \tag{6.2.3a}$$

$$fu_I = 0, \tag{6.2.3b}$$

$$w_{Iz} = 0. \tag{6.2.3c}$$

(where $\zeta$ is the free-surface displacement and $g$ is the acceleration due to gravity). Keep in mind that, on the inner shelf, there is not actually a spatially distinct inviscid interior, although it remains valid (for this linear problem where $A$ does not depend on the actual velocity) to break out a component having inviscid dynamics. On the inner shelf, then, it is best to think of $v_I$ as being a measure of the cross-shelf pressure gradient rather than an actual current. The boundary layer (viscous) component of the flow is governed by

$$-fv_E = (Au_{Ez})_z, \tag{6.2.4a}$$

$$fu_E = (Av_{Ez})_z. \tag{6.2.4b}$$

The boundary conditions are

$$\rho_0 A(u_{Ez}, v_{Ez}) = (\tau_0^x, \tau_0^y) \qquad \text{at } z = 0 \tag{6.2.5a}$$

and, at the bottom,

$$u_I + u_E = u_E = 0, \tag{6.2.5b}$$

$$v_I + v_E = 0 \qquad \text{at } z = -h(x) \tag{6.2.5c}$$

where $(\tau_0^x, \tau_0^y)$ is the imposed surface wind stress.

With a uniform $A$, it is straightforward, but complicated, to solve the boundary layer problem without making any assumptions about the Ekman scale depth

$$\delta_E = \left(\frac{2A}{|f|}\right)^{1/2}. \tag{6.2.6}$$

relative to the water depth $h(x)$ (Ekman, 1905). Using a constant eddy viscosity and no-slip boundary conditions (6.2.5b, c) is not particularly realistic, but the results remain illustrative. The solutions are encapsulated in terms of the total Ekman transport $(U_E, V_E)$ where, for example,

$$U_E = \int_{-h}^{0} u_E \, dz \tag{6.2.7}$$

so that (in the northern hemisphere), allowing $u_I \neq 0$ for now,

$$U_E = (\rho_0 f)^{-1}(1 - S_1)\tau_0^y - (\rho_0 f)^{-1} S_2 \tau_0^x - \frac{v_I \delta_E (T_1 - T_2)}{2} - \frac{u_I \delta_E (T_1 + T_2)}{2} \tag{6.2.8a}$$

$$V_E = (\rho_0 f)^{-1}(S_1 - 1)\tau_0^x - (\rho_0 f)^{-1} S_2 \tau_0^y + \frac{u_I \delta_E (T_1 - T_2)}{2} - \frac{v_I \delta_E (T_1 + T_2)}{2} \tag{6.2.8b}$$

These expressions depend upon the functions (Figure 6.2)

$$S_1 = \gamma^{-1} \cos(\xi) \cosh(\xi), \tag{6.2.9a}$$

$$S_2 = -\gamma^{-1} \sin(\xi) \sinh(\xi), \tag{6.2.9b}$$

$$T_1 = \gamma^{-1} \sinh(\xi) \cosh(\xi), \tag{6.2.9c}$$

$$T_2 = \gamma^{-1} \sin(\xi) \cos(\xi), \tag{6.2.9d}$$

$$\gamma = [\cos(\xi) \cosh(\xi)]^2 + [\sin(\xi) \sinh(\xi)]^2, \tag{6.2.9e}$$

where

$$\xi(x) = h(x)/\delta_E. \tag{6.2.9f}$$

(Note that $\xi^{-2}$ is effectively a local Ekman number.)

Needless to say, expressions (6.2.8) and (6.2.9) are not very illuminating. Some clarity can be obtained by considering the limits of large and small $\xi$. In water that is deep compared with the Ekman scale depth, $\xi \gg 1$, the limiting forms for (6.2.9) are

$$T_1 \approx 1, \tag{6.2.10a}$$

$$T_2, S_1, S_2 \approx 0, \tag{6.2.10b}$$

so that the familiar (i.e., 3.2.11 and 3.3.4) form results:

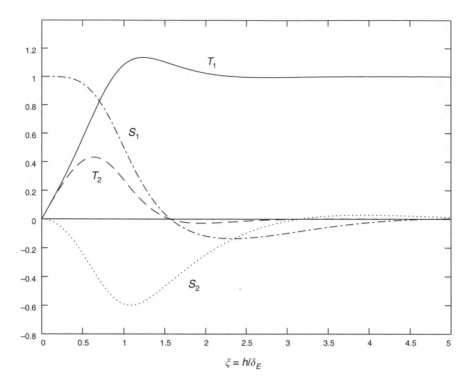

**FIGURE 6.2.** Frictional transport functions (6.2.9a–6.2.9d).

$$U_E = (\rho_0 f)^{-1} \tau_0^y - v_I \delta_E / 2 - u_I \delta_E / 2, \tag{6.2.11a}$$

$$V_E = (\rho_0 f)^{-1} \tau_0^x + u_I \delta_E / 2 - v_I \delta_E / 2, \tag{6.2.11b}$$

On the other hand, when the water is very shallow, $\xi \to 0$,

$$T_1 \approx T_2 \approx \xi, \tag{6.2.12a}$$

$$S_1 \approx 1, \tag{6.2.12b}$$

$$S_2 \approx -\xi^2, \tag{6.2.12c}$$

so that, returning to the alongshore-uniform case where $u_I = 0$,

$$U_E \approx (\rho_0 f)^{-1} \tau_0^x \xi^2 - u_I h = (\rho_0 f)^{-1} \tau_0^x \xi^2 = (2\rho_0 A)^{-1} h^2 \tau_0^x, \tag{6.2.12d}$$

$$V_E \approx (\rho_0 f)^{-1} \tau_0^y \xi^2 - v_I h. \tag{6.2.12e}$$

The last equation then implies that

$$\int_{-h}^{0} v \, dz = V_E + v_I h \approx (\rho_0 f)^{-1} \tau_0^y \xi^2 = (2\rho_0 A)^{-1} h^2 \tau_0^y. \tag{6.2.12f}$$

Thus, in shallow water, the interior alongshore velocity is neutralized, and overall velocity is very weak, proportional to $h^2$. The total transport (e.g., 6.2.12d, f) is independent of rotation, hence strictly downwind. Altogether, averaging over turbulent fluctuations, the ocean becomes nearly motionless as the depth goes to zero: the surface wind stress is passed directly to the ocean bottom without greatly affecting the vanishingly thin ocean. This stress transfer through a motionless ocean may, appropriately, strike the reader as bizarre; it is a consequence of the bottom stress being dependent on the velocity gradient rather than on the velocity itself, as is appropriate close to the bottom (section 3.6). In any case, a colleague once noted, "the ocean does not have a side boundary: the top and the bottom simply come together."

The most interesting aspect of this problem, then, is the intermediate range where the water depth is the same magnitude as the Ekman layer scale thickness, that is, where $\xi = O(1)$. This marks the transition zone from where there is a distinct inviscid interior ($\xi \gg 1$) to where turbulent stresses dominate over rotation. In between, the turbulent "boundary layer" fills the entire water column, and the surface and bottom Ekman layers cease to be distinct; rather, they merge and ultimately cancel as $\xi$ decreases. This nearshore shutdown, in turn, implies a *vertical* transport within this region if, as in the coastal upwelling example, opposing inflow and outflow occur in distinct boundary layers where the inner shelf joins the midshelf.

The results are illustrated by a sample calculation (Figure 6.3). Alongshore variations are neglected, and $u_I = 0$ everywhere in a steady, two-dimensional $(x, z)$ problem (6.2.3b). Thus, using the depth integral of the continuity equation (6.2.1c) and the coastal boundary condition, $U_E = 0$ everywhere. Given these conditions, (6.2.8a) becomes

$$0 = (\rho_0 f)^{-1}(1 - S_1)\tau_0^y - (\rho_0 f)^{-1} S_2 \tau_0^x - \frac{v_I \delta_E (T_1 - T_2)}{2}. \tag{6.2.13}$$

Now, consider the case of $\tau_0^x = 0$, and $\tau_0^y$ being constant and uniform. Equation (6.2.13) then requires that

$$v_I = \frac{2\tau_0^y (1 - S_1)}{\rho_0 f \delta_E (T_1 - T_2)}, \tag{6.2.14}$$

which, through (6.2.3a), also determines the sea level slope everywhere. The overall results are shown as the solid curves in Figure 6.3. $Q_E$ is the net offshore transport over depths shallower than $z = -h(x)/2$, normalized by the far-field Ekman transport $U_{E0} = (\rho_0 f)^{-1} \tau_0^y$. The vertical velocity $w_M$ is calculated at mid-depth (i.e., at $z = -h(x)/2$) and expresses the closure of a transport loop involving, for example, deep onshore flow, upwelling in roughly the range $1 < h/\delta_E < 5$, and shallow offshore Ekman transport. $Q_E$ and $v_I$ reach roughly their ultimate, offshore values by around $h/\delta_E \geq 4$. Indeed, for these larger depths, the solution is exactly what would be expected based on simpler models (section 4.3). What is new here is the demonstration of how wind-driven cross-shelf flow weakens in shallow water and, consequently, how friction supports the required inner shelf vertical transport.

Now, consider the response to a uniform, steady *cross-shelf* wind stress. Once again, $u_I = U_E = 0$, so that (6.2.13) becomes

$$0 = -(\rho_0 f)^{-1} S_2 \tau_0^x - \frac{v_I \delta_E (T_1 - T_2)}{2}, \tag{6.2.15}$$

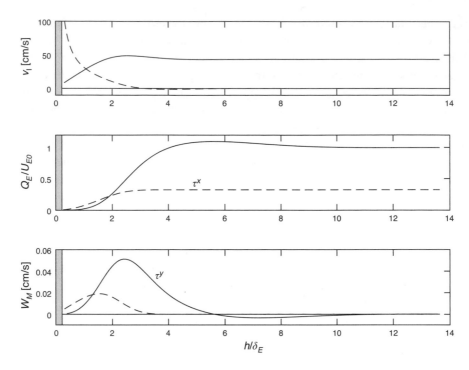

**FIGURE 6.3.** Spatial changes over the transition from inner- to midshelf (small to large $h/\delta_E$): results from an idealized steady two-dimensional model with constant eddy viscosity. Upper panel: interior alongshore velocity (actually, a measure of the pressure gradient); Middle panel: offshore transport normalized by deep-water Ekman transport $U_{E0} = |\tau_0|/(\rho_0 f)$; Lower panel: mid-depth vertical velocity for a 0.1 N/m² wind stress in the alongshore direction (solid curves) and offshore direction (dashed curves). The horizontal axis is the water depth normalized by the Ekman scale depth. The water depth is $h(x) = 1\,\text{m} + 5\times10^{-3}\,x$, $f = 1\times10^{-4}\,1/\text{s}$, and $A = 1\times10^{-3}\,\text{m}^2/\text{s}$. Thus, the Ekman scale depth $\delta_E = 4.5\,\text{m}$. The shaded area corresponds to $h < 1\,\text{m}$, where there is no ocean.

and

$$v_I = \frac{-2\tau_0^x S_2}{\rho_0 f \delta_E (T_1 - T_2)}. \tag{6.2.16}$$

The cross-shelf pressure gradient (expressed as the interior alongshore velocity) is confined to the inner shelf (Figure 6.3, dashed lines) and consists of a modest set-down of near-shore sea level in response to an offshore wind: less than 0.01 m for this idealized example using $\tau_0^x = 0.1$ N/m² and $A = 1\times10^{-3}$ m²/s. Far from shore ($h/\delta_E \gg 1$), the total offshore Ekman transport $U_E$ goes to zero, but near the surface offshore flow is balanced, just below and still within the surface Ekman layer, by a compensating onshore flow. This is a consequence of the directional rotation that defines the Ekman spiral. Thus, $Q_E$ (defined now by a division within the surface Ekman layer rather than by some greater depth) is nonzero but weaker than it would be for an equivalent alongshore wind stress. This pattern extends well into the inner shelf (e.g., Lentz and Fewings, 2012), so there is a shallow upwelling cell, again closed via a vertical transport (for $\xi < 3$) in the frictionally dominated inner shelf.

Given a better understanding of the role of the inner shelf, one can now reconsider flow in water deep enough that the surface and bottom boundary layers are distinct (e.g., 6.2.11). Very often, to avoid the complexities of the inner shelf region, theories assume an arbitrary coastal wall at some isobath, perhaps 10–30 m. It is assumed that the total cross-shelf transport at the wall adds to zero, even if alongshore variability is allowed, so that $u_I \neq 0$. The assumptions are that alongshore transport divergence is negligible closer to the physical shore and that vertical motions occur in shallower water (inshore of the wall) but that the details do not affect flow farther offshore. Mitchum and Clarke (1986a) carefully considered linear barotropic shelf models using the coastal long-wave approximation; that is, they allowed both large-scale alongshore dependence and moderate time dependence. They solved the problem both resolving the inner shelf and using a vertical coastal wall. They concluded that using a wall gives reasonable results over the mid- to outer shelf when the wall is located where $h(x) = 3\delta_E$. This finding validates the way the simple two-dimensional models of chapter 4 skirt the inner shelf, but it is not obvious that their result holds when alongshore scales are shorter. Another approach, which does not require the use of the coastal long-wave approximation, to defining an inshore boundary location is to use (4.1.1) to estimate the width of the region where flow is two-dimensional.

To this point a constant eddy viscosity, uniform in space, has been applied. Lentz (1995) treated the steady inner shelf boundary layer problem using a variety of different steady vertical $A$ profiles, all of which are more realistic than a constant. Specifically, he considered several structures where the eddy viscosity vanishes near the upper and lower boundaries, as is generally expected in a turbulent flow (section 3.6). Qualitatively, nothing changes in the cross-shelf circulation, but there are clear quantitative differences among the different examples.

Turbulence, hence eddy viscosity, in fact varies with, among other factors, local Richardson number

$$Ri = \frac{-g\rho_z}{\rho_0 v_z^2} \tag{6.2.17}$$

and is thus highly variable in space and time. Some insight can be gained from the well-resolved two-dimensional numerical models described in section 4.6, because they include sophisticated turbulence closure schemes that account for space and time variability in mixing and stratification. Earlier discussion dealt with the development of midshelf interior and boundary layer flows, but now consider the shallower waters. For example, numerical experiments with steady alongshore winds (Austin and Lentz, 2002) showed dramatically different inner shelf results depending on wind direction (Figure 6.4). During upwelling, deep, dense water is carried into the inner shelf and is then transported offshore by the surface Ekman transport. The influx of dense water tends to keep at least some of the inner shelf stably stratified and thus only moderately turbulent. Water passes through the inner shelf in a way not too different than in the constant eddy viscosity model (Figure 6.3, solid lines). As denser, upwelled water is carried offshore, it initially passes above lighter ambient water, and so near-surface mixing is enhanced because of the locally

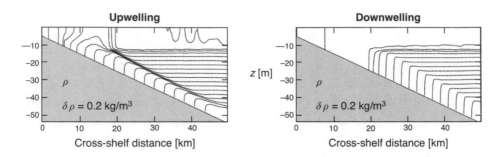

**FIGURE 6.4.** Density sections for idealized two-dimensional model runs forced by an upwelling-favorable alongshore wind stress (left panel) and a downwelling-favorable wind stress (right panel). Other than the direction of the wind stress, the two runs are identical. Adapted from Austin and Lentz (2002).

unstable stratification. Farther offshore, near-surface horizontal density gradients weaken, and mixing is effectively a one-dimensional process. In contrast, with downwelling-favorable winds, light near-surface waters are transported onshore, where they effectively flood the inner shelf. Most of the water found on the inner shelf thus comes from the spatially uniform near-surface region offshore (Figure 6.4, right panel). Thus, density stratification here is almost absent, and turbulence is very well developed. The turbulence level, in turn, strongly inhibits inner shelf flow, so that all downwelling is concentrated near the boundary between stratified and mixed waters, at $x = 20$ km. These model results are qualitatively consistent with many coastal observations.

The important point here is that, while constant eddy viscosity results are illuminating, the actual inner shelf is more complex because of the sensitivity of mixing to stratification and flow conditions. The inner shelf thus exhibits substantial nonlinearity. For example, Horwitz and Lentz (2016) used a model and multiyear inner shelf observatory records to study current response to cross-shelf wind stress under initially stratified conditions where the alongshore wind stress is nonzero. They found that when the accompanying alongshore wind stress component is upwelling-favorable, the inner shelf is somewhat stratified, so that both alongshore and cross-shelf wind stresses are effective at flushing the inner shelf. In contrast, when the accompanying alongshore component is downwelling-favorable, the inner shelf is well mixed and highly turbulent (and hence has a large eddy viscosity in their accompanying model). Inner shelf velocity is then inefficiently forced by alongshore winds, although the cross-shelf wind stress remains effective. Combining these results and considering all wind directions, Horwitz and Lentz found there is a bias favoring offshore near-surface transport on the inner shelf.

## 6.3.    Nonhydrostatic Gravity Waves

Before examining how surface waves drive flow on the inner shelf, it is useful to review the properties of short (nonhydrostatic) gravity waves. These are the familiar, atmospherically forced (e.g., Sullivan and McWilliams, 2010), high-frequency (periods of a few tens of seconds or often less) waves that help make shipboard life interesting. Because of their high frequency, Earth's rotation can be ignored for the time being, so the two-dimensional linear governing equations with no density stratification are

$$\tilde{u}_t = -\frac{1}{\rho_0}\tilde{p}_x, \tag{6.3.1a}$$

$$\tilde{w}_t = -\frac{1}{\rho_0}\tilde{p}_z, \tag{6.3.1b}$$

$$0 = \tilde{u}_x + \tilde{w}_z, \tag{6.3.1c}$$

with a surface boundary condition that

$$\tilde{w} = \tilde{\varsigma}_t = \frac{1}{g\rho_0}\tilde{p}_t \qquad \text{at } z = 0, \tag{6.3.1d}$$

and at the flat bottom

$$\tilde{w} = 0 \qquad \text{at } z = -h. \tag{6.3.1e}$$

The tilde denotes variables associated with the gravity wave. Combining the first three equations (6.3.1a, b, c),

$$0 = \tilde{p}_{xx} + \tilde{p}_{zz}, \tag{6.3.2a}$$

subject to

$$\tilde{p}_{tt} = -g\tilde{p}_z \qquad \text{at } z = 0, \tag{6.3.2b}$$

and

$$\tilde{p}_z = 0 \qquad \text{at } z = -h. \tag{6.3.2c}$$

A solution (Figure 6.5) to this system is then

$$\tilde{p} = C\cosh[k(z+h)]\cos(kx - \omega t), \tag{6.3.3a}$$

$$\tilde{u} = \frac{Ck}{\omega\rho_0}\cosh[k(z+h)]\cos(kx - \omega t), \tag{6.3.3b}$$

$$\tilde{w} = \frac{Ck}{\omega\rho_0}\sinh[k(z+h)]\sin(kx - \omega t), \tag{6.3.3c}$$

where $C$ is a constant amplitude, and

$$\omega^2 = gk\tanh(kh). \tag{6.3.3d}$$

Note that frequency is not a linear function of wavenumber, so these waves can be highly dispersive (see section 5.2) and thus can evolve into a range of sinusoids, even given a compact initial state that is not sinusoidal.

In water that is deep compared with a wavelength, $kh \gg 1$, the wave's structure is confined near the surface, so that

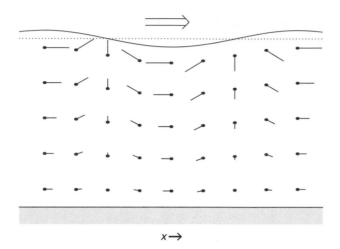

**FIGURE 6.5.** Schematic of a gravity wave propagating toward the right, with instantaneous velocity shown as vectors, each with its origin at a black dot. The arbitrary velocity scales are not the same for vertical and horizontal directions. The arrow above the surface indicates the direction of propagation.

$$\tilde{p} \cong Ge^{kz}\cos(kx - \omega t),\tag{6.3.4a}$$

and

$$\omega^2 \cong gk.\tag{6.3.4b}$$

For ocean swell with a period of 10 s, this gives a wavelength of about 150 m, and the wave has greatly attenuated amplitude below about 50 m. Thus, as such a wave propagates toward the coast, it does not begin to "feel" the bottom and take the form (6.3.3) until on the shelf, with water depth less than about 50 m. In contrast, for waves that are long relative to the water depth, $kh \ll 1$, hydrostatic conditions (section 5.2) apply, and

$$\tilde{p} = C\cos(kx - \omega t).\tag{6.3.5a}$$

Because $\tanh(kh) \approx kh$ in this limit,

$$\omega^2 \cong ghk^2,\tag{6.3.5b}$$

and the long waves are nondispersive (phase speed $\sqrt{gh}$), traveling faster as $h$ increases.

Thus, as a wave propagates from deep water onto the shelf, $h$ decreases, and its structure changes from surface trapped (6.3.4a) to more distributed vertically (6.3.3a), and the wave's group velocity becomes dependent on $h$. These changes alone can account for substantial modification of the wave field as it approaches land. In addition, when the water is shallow enough, the wave's surface displacement amplitude $\tilde{\zeta}$ becomes comparable to the water depth $h$; that is, the wave's dynamics become nonlinear. Given all these effects, an incoming wave often steepens and breaks as it passes into shallow water, generating turbulence and causing wave dissipation. The surf zone is defined by the prevalence of these breaking waves. The entire process of cross-shelf wave propagation is complex and fascinating; it deserves the fuller treatment provided by Dean and Dalrymple (1992), for example.

Based on (6.3.3b), it is clear that the Eulerian mean velocity associated with a wave is zero. In other words, if an average is calculated at a strictly fixed location below the excursion range of the free surface, there is no mean flow, $\langle \tilde{u} \rangle = 0$, where the angle brackets denote a time average over multiple wave periods. However, a water particle in a wave field does not stay at an exactly fixed location but, rather, moves both horizontally and vertically during the course of a cycle. Thus, the total Lagrangian mean velocity (i.e., the net velocity following a particle) is determined at least partly by the fact that a particle experiences different velocity ellipse amplitudes at different locations as it moves about. Thus, expanding velocity as $q = q_0 + q_x \xi^x + q_z \xi^y$ (where $\xi^x$ and $\xi^y$ are particle displacements), the Lagrangian mean velocity can be expressed to a good approximation as

$$\langle u_L \rangle = \langle \tilde{u} \rangle + \langle \xi^x \tilde{u}_x \rangle + \langle \xi^z \tilde{u}_z \rangle, \tag{6.3.6a}$$

$$\langle w_L \rangle = \langle \tilde{w} \rangle + \langle \xi^x \tilde{w}_x \rangle + \langle \xi^z \tilde{w}_z \rangle, \tag{6.3.6b}$$

with

$$\xi^x = \int \tilde{u} \, dt, \tag{6.3.6c}$$

$$\xi^z = \int \tilde{w} \, dt. \tag{6.3.6d}$$

The averaged product terms (e.g., $\langle \xi^x \tilde{u}_x \rangle$) collectively constitute the *Stokes* (1847) *drift*, and in this notation, $\langle \tilde{u} \rangle$ is the Eulerian (fixed-point) mean velocity. A Stokes drift is possible when the wave's particle velocity involves spatial gradients (vertical, in this case) and where particle excursions across that gradient occur. The result is that each water parcel follows a looping (as opposed to closed, elliptical) trajectory, experiencing in Figure 6.5 stronger positive (for propagation in the positive $x$ direction) horizontal velocities when displaced upward, and weaker negative velocity when displaced downward. Thus, on average, the particle feels stronger positive than negative velocity over the course of a period, so it undergoes a net positive displacement. Given (6.3.3), the Stokes drift is

$$u_S = \frac{\zeta_0^2 \, k\omega \, \cosh[2k(z+h)]}{2[\sinh(kh)]^2}, \tag{6.3.7a}$$

$$w_s = 0, \tag{6.3.7b}$$

where $\zeta_0$ is the amplitude of the free-surface elevation[1]

$$\zeta_0 = C(g\rho_0)^{-1} \cosh(kh). \tag{6.3.7c}$$

---

[1] Very often, observed surface wave amplitude is expressed in terms of the "significant wave height" $H_s$. This is defined as the average height of the largest one-third of waves present. While this definition sounds obscure, it corresponds well to what one's eyes perceive. It is often reasonable to estimate $\zeta_0$ as $H_s/4$ (e.g., Lentz and Fewings, 2012) and to pick a dominant wave frequency.

The Stokes drift is thus in the direction of wave propagation and is strongest near the surface; in deep water ($kh \gg 1$), (6.3.7a) reduces to

$$u_S \cong \zeta_0^2 \, k\omega \, e^{2kz}. \tag{6.3.7d}$$

Note, too, that, with all else being the same, (6.3.7a) shows that the Stokes drift tends to grow stronger as water becomes shallower (assuming that $C$ is constant, which is not probable).

Gravity waves, of course, transport momentum, and this can be a primary driver of Eulerian motion on the inner shelf, and certainly in the surf zone. Specifically, the momentum transport by a wave is given by its radiation stress (Longuet-Higgins and Stewart, 1962, 1964)

$$S^{xx} = \left\langle \int_{-h}^{\zeta} (\rho_0 \tilde{u}^2 + \tilde{p}) \, dz \right\rangle, \tag{6.3.8a}$$

$$S^{xy} = \left\langle \int_{-h}^{\zeta} \rho_0 \tilde{u}\tilde{v} \, dz \right\rangle, \tag{6.3.8b}$$

where the angle brackets represent a time average, and ($S^{xx}$, $S^{xy}$) represent the mean $x$-directed flux of $x$ (offshore) and of $y$ (alongshore) momentum, respectively. It is assumed that the hydrostatic component of pressure, $-\rho_0 g z$, has been removed from $\tilde{p}$ (as is the case in eqn. 6.3.3a). Note that since $\tilde{\zeta}$ (the wave's free-surface displacement) appears in the integral bounds, these are potentially complicated expressions in practice. For linear, monochromatic waves, the radiation stress can be written using (6.3.3) and (6.3.8) (Mei, 1983, his chapter 10) as

$$S^{xx} = E_0 \left[ \frac{c_g}{c_p}(\cos^2\theta + 1) - \frac{1}{2} \right], \tag{6.3.9a}$$

$$S^{xy} = E_0 \frac{c_g}{c_p}\cos\theta\sin\theta, \tag{6.3.9b}$$

where

$$E_0 = \rho_0 g \zeta_0^2, \tag{6.3.9c}$$

and $\theta$ is the direction of wave propagation (phase and group velocity are both parallel to the wavenumber vector) measured counterclockwise relative to the $x$ (offshore) direction. The wave's phase velocity and group velocity (see section 5.2 for definitions and explanations) are, respectively,

$$c_p = \frac{\omega}{\kappa} = \frac{1}{\kappa}\sqrt{g\kappa \, \tanh(\kappa h)}, \tag{6.3.9d}$$

and

$$c_g = \frac{\partial \omega}{\partial \kappa} = \frac{1}{2}c_p \left[ 1 + \frac{2\kappa h}{\sinh(2\kappa h)} \right], \tag{6.3.9e}$$

where $\kappa = \sqrt{k^2 + l^2}$ is in the direction of wave propagation. Alternatively, the ability for waves to transfer mean momentum can be expressed in terms of a vortex force

(McWilliams et al., 2004), which is essentially equivalent under realistic inner shelf conditions. The vortex force formalism offers more physical clarity in some regards but involves difficulties when it comes to wave breaking (Lane et al., 2007).

It is not the momentum fluxes themselves that are of greatest interest but, rather, the divergence of the fluxes. After all, the mean momentum that is left in some volume is the difference between the inward and outward fluxes. Even in the simplest of circumstances (directly onshore propagation with no dissipation), $S^{xx}$ changes owing to the wave's changing structure as the water becomes shallower. In reality, effects such as steepening, frictional dissipation, scattering, and refraction all affect gravity waves as they propagate into shallower water (e.g., Ardhuin et al., 2003). Once water is shallow enough (3 m is a number often given, but in reality, this depth varies considerably depending on wave amplitude), wave breaking becomes important, dissipating a large amount of wave energy, and so the radiation stress *convergence* becomes large indeed. To visualize this point, imagine a volume where waves propagate in but dissipate inside. There is a momentum flux in the one side but no waves, hence no flux, out the other.

## 6.4.    Wind and Wave Forcing

Two important types of inner shelf forcing—winds and waves—have been considered thus far, and their relative importance is treated next. It is important to remember that there are other potentially important forcing mechanisms. For example, frequently buoyancy forcing (chapter 8) or tidal rectification (section 7.4) is an important driver as well. The approach here is to use observations to illustrate balances typical of the inner shelf and its neighbors, and then to examine a few of the limiting behaviors.

Lentz et al. (1999) reported on a comprehensive 4-month set of fixed-site current measurements off North Carolina that ranged from the surf zone (4 m water depth) out 16 km offshore to the 26 m isobath. The observations encompassed late summer into the late fall. Under these conditions, the outermost mooring is considered typical of the mid-shelf as opposed to the inner shelf. The data allow estimation of most of the terms in the depth-averaged, low-pass-filtered (anything with a period shorter than about 38 hours is removed) momentum equations

$$\frac{\partial \bar{u}}{\partial t} + \frac{1}{h}\frac{\partial}{\partial x}\int_{-h}^{0} u^2 \, dz + \frac{1}{h}\frac{\partial}{\partial y}\int_{-h}^{0} uv dz - f\bar{v} = -\frac{1}{\rho_0}\frac{\partial \bar{P}}{\partial x} + \frac{\tau_0^x}{h\rho_0} - \frac{\tau_B^x}{h\rho_0} - \frac{1}{h\rho_0}\frac{\partial S^{xx}}{\partial x},$$

$$(6.4.1a)$$

$$\frac{\partial \bar{v}}{\partial t} + \frac{1}{h}\frac{\partial}{\partial x}\int_{-h}^{0} uv \, dz + \frac{1}{h}\frac{\partial}{\partial y}\int_{-h}^{0} v^2 dz + f\bar{u} = -\frac{1}{\rho_0}\frac{\partial \bar{P}}{\partial y} + \frac{\tau_0^y}{h\rho_0} - \frac{\tau_B^y}{h\rho_0} - \frac{1}{h\rho_0}\frac{\partial S^{xy}}{\partial x},$$

$$(6.4.1b)$$

where the overbar symbol represents a vertical average, $\bar{P}$ is vertically averaged pressure, and $(\tau_B^x, \tau_B^y)$ is the bottom stress. Nonlinear terms are computed with the original data and then smoothed, so that the residual effects of higher frequency motions are included.

The standard deviation is used as a measure of the importance of the different terms in time-dependent balances. The vertical integral terms account for variability, such as tides, only on time scales longer than those of surface waves, since the wave forcing is accounted for in the radiation stress terms.

The midshelf (26 m depth) cross-shelf momentum balance is well approximated simply by geostrophy

$$f\bar{v} = \frac{1}{\rho_0}\frac{\partial \bar{P}}{\partial x}, \tag{6.4.2a}$$

while the alongshore equation there is

$$\frac{\partial \bar{v}}{\partial t} + f\bar{u} = -\frac{1}{\rho_0}\frac{\partial \bar{P}}{\partial y} + \frac{\tau_0^y}{h\rho_0} - \frac{\tau_B^y}{h\rho_0}. \tag{6.4.2b}$$

These balances are typical of what is expected on these time scales on the midshelf (e.g., section 5.7) and are consistent with the existence of a distinct "interior" along with surface and bottom boundary layers. Note that the gradients of radiation stress become weak and then vanish in deep water where $kh \gg 1$.

On the inner shelf (e.g., at 8 m depth), the balances are well-approximated by

$$-f\bar{v} = -\frac{1}{\rho_0}\frac{\partial \bar{P}}{\partial x} - \frac{1}{h\rho_0}\frac{\partial S^{xx}}{\partial x}, \tag{6.4.3a}$$

$$0 = -\frac{1}{\rho_0}\frac{\partial \bar{P}}{\partial y} + \frac{\tau_0^y}{h\rho_0} - \frac{\tau_B^y}{h\rho_0}. \tag{6.4.3b}$$

Although alongshore velocity magnitudes are similar to those at midshelf, $v$ is no longer geostrophically balanced, because wave radiation stress divergence now enters. The alongshore momentum equation is simplified because the acceleration and Coriolis terms become relatively weaker, while both the alongshore surface and bottom stresses remain important, as expected from section 6.2. The important point here is that both wind and wave forcing are important in this depth range.

Finally, since the surf zone is defined by the presence of breaking waves, its bounds fluctuate considerably, depending on the amplitude and scale of the incoming surface waves. When the 4 and 8 m sites are within the surf zone, their momentum balances are roughly

$$0 = -\frac{1}{\rho_0}\frac{\partial \bar{P}}{\partial x} - \frac{1}{h\rho_0}\frac{\partial S^{xx}}{\partial x}, \tag{6.4.4a}$$

and

$$0 = -\frac{\tau_B^y}{h\rho_0} - \frac{1}{h\rho_0}\frac{\partial S^{xy}}{\partial x}. \tag{6.4.4b}$$

The former balance shows that a free-surface tilt (pressure gradient) balances the incoming, cross-shelf radiation stress divergence. This balance is called *wave setup* (Bowen et al., 1968), and typical surface deflections can range up to tens of centimeters (e.g., Lentz and Raubenheimer, 1999). The second balance expresses the ability of the onshore wave transport of alongshore momentum to drive alongshore currents, which lead to a bottom stress, in the surf zone (Thornton and Guza, 1986). Typically, radiation stress variations can be an order of magnitude greater in the surf zone (where the dissipation tied to breaking waves causes large radiation stress gradients) than on the inner shelf (where radiation stress divergence is associated with less drastic effects such as gradual changes owing to waves adiabatically shoaling). Note that in the surf zone neither component of the surface wind stress plays a role; it is not that the wind stress is small but, rather, that the radiation stress divergence is large.

## 6.5.    Cross-Shelf Transport on the Inner Shelf

On time scales longer than those of tides, cross-shelf currents over the inner shelf can be quite small. For example, representative mean currents in 12 m of water south of Martha's Vineyard are perhaps 0.02 m/s. However, the cross-sectional area of water inshore of this site is about $9 \times 10^3$ m$^2$, so the flushing time (area divided by velocity and depth) is less than a half day. Thus, although the current is weak, the inshore area is small, so flushing is rapid. As a first approximation, it is often assumed that the net cross-shelf transport over the inner shelf sums to zero locally, that is, that the onshore and offshore transports balance, and the divergence of alongshore flow is negligible. If this is true, then the various types of inner shelf cross-shelf volume fluxes have to balance.

One form of onshore transport, already mentioned, is the Stokes drift. Since surface waves propagate primarily toward the coast (reflected waves, although present, tend to be weaker due to dissipation), the Stokes drift (6.3.7) is generally toward the coast. Although this net motion cannot be directly measured by a current meter (which can ideally measure only Eulerian currents), it does represent an onshore transport that needs to be balanced. In addition, frictional effects in a thin wave bottom boundary layer give rise to a mean flow, called *streaming*, very near the bottom. In the simplest case, with molecular viscosity (Longuet-Higgins, 1953), stresses in the wave bottom boundary layer $(O(v/\omega)^{1/2}$ thick, where $v$ is the molecular viscosity) allow $\tilde{u}$ and $\tilde{w}$ to have components in phase with each other, creating a Reynolds stress $\langle \tilde{u}\tilde{w} \rangle$ and thus a mean flow in the direction of wave propagation. When models allow for a turbulent wave boundary layer, complexities arise, and even the direction of the mean streaming is not a given (e.g., Trowbridge and Lentz, 2018). Although realistic numerical models predict an active layer that is perhaps a few centimeters thick (Uchyama et al., 2010) with velocity $O(0.1$ m/s), field observations of streaming have been elusive. However, model results (e.g., Wang et al., 2020) demonstrate that the inclusion of streaming in inner shelf/surf zone models leads to pronounced circulation differences in the remainder of the water column.

Another form of wave-driven cross-shelf flow is due to the Stokes-Coriolis acceleration (Hasselmann, 1970). In this case, the Coriolis force acts on gravity waves. Despite the waves' high frequency and the consequent weakness of the Coriolis effect over a single

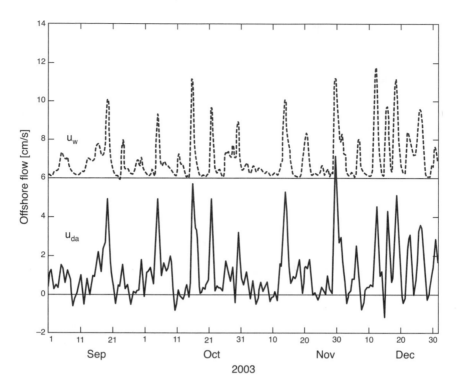

**FIGURE 6.6.** Observed depth-averaged Eulerian *off*shore velocity at the 12 m isobath south of Martha's Vineyard (solid line) and estimated depth-averaged *on*shore Stokes drift estimated at the same location (dashed curve), with 6 cm/s arbitrarily added to the Stokes drift estimates to separate the curves in this plot. The similarity of the two curves indicates that the two transports tend to balance. Adapted from Lentz et al. (2008).

wave period, the cumulative forcing over many periods leads to a substantial Eulerian mean inner shelf current $O(0.01$ m/s) when frictional effects are negligible. This flow is in the direction counter to that of wave propagation (i.e., offshore) and has a vertical structure similar but opposed to that of the Stokes drift. In contrast, when dissipation plays a stronger role, rotational effects become less important, and the rectified mean offshore flow has a more parabolic profile, as is often found in the surf zone (e.g., Reniers et al., 2004). Using inner shelf measurements (depth of order 12 m) from three different locations, Lentz et al. (2008) estimated the Stokes drift and measured the depth-averaged Eulerian flow (Figure 6.6): the two generally balanced strikingly well. Further, they used information about the observed wave field, along with assumptions about the vertical eddy viscosity, to force a skillful model of the mean cross-shelf currents. They found that the Stokes-Coriolis forcing is the most important mechanism driving the offshore Eulerian mean flow that balances the Stokes drift. Very long time series from a permanent observatory then allowed them to calculate vertical flow profiles under a range of different forcing conditions, and they found that with weak winds the Stokes-Coriolis flow tends to balance the Stokes drift depth-by-depth. More recent models of detailed turbulent flow (Pearson, 2018), however, have raised questions about how turbulent eddies, interacting with surface waves, can drive a mean flow even in the absence of Earth's rotation. Clearly, there is still more to learn about transport through the inner shelf.

## 6.6.   Perspective

The inner shelf is a dynamically defined region, so its bounds can vary on time scales of hours to days and longer. It is characterized by the merger of the surface and bottom turbulent boundary layers, and by the increasing importance of wave forcing as the water grows shallower. This chapter deals only with wind and wave forcing, which are undeniably important, but effects associated with buoyancy currents and tidal forcing, discussed in other chapters, are often important here as well.

The discussion has consistently ignored alongshore variability and, clearly, in some situations (e.g., Figure 6.6) this is a reasonable approximation. However, there is increasing evidence that alongshore variability is sometimes important, especially when data are not averaged over time intervals of days or longer. For example, high-frequency-radar surface-current maps (Kirincich et al., 2013) demonstrated that both tidal and mean currents have substantial alongshore variations over the inner shelf south of Martha's Vineyard. Further, Kirincich (2016) showed that the inner shelf here is often populated by small (2–5 km), short-lived (hours) eddies that appear to be at least partially generated by tidal flow across topography. Much of the mean flow structure here is determined by tidal rectification (Ganju et al., 2011; see also section 7.4), which implies a strong tendency for mean velocity to follow isobaths even when they are not parallel to the coast. It is interesting that these undeniably three-dimensional phenomena are so clearly present in the same general area where fixed-point measurements (such as in Figure 6.6), when smoothed over a couple days' time scale, appear two-dimensional. Evidently, the long-term observatory resides in a spot where alongshore variability is unusually weak.

The finding of three-dimensional, time-dependent features is hardly unique to the Martha's Vineyard example. Along a smooth beach, it appears that a major exchange pathway between the surf zone and the inner shelf consists of intense (ultimately wave-driven) *rip currents*: short-lived, energetic offshore jets typically 10–200 m wide (e.g., Dalrymple et al., 2011; Haly-Rosendahl et al., 2015; Moulton et al., 2021). The inner shelf often contains miniature eddies that are associated with instabilities of alongshore flows (e.g., section 9.10) or especially with irregularities in the incoming surface or internal wave fields (e.g., Wang, et al., 2020; McSweeney et al., 2020). Further, the inner shelf is also home to oblique (i.e., not shore-parallel) fronts that can extend from the inner to midshelf (e.g., Wu et al., 2021). Our knowledge of the inner shelf is conditioned by a very understandable bias toward making measurements along shorelines that consist of straight, sandy beaches. Alongshore variability associated with irregular shorelines and rocky headlands has received little attention until recently.

All considered, there is a clear need to continue dealing systematically with along-shore variability, and its integrated effects, over the inner shelf. While some features, such as eddies, tend to average out over time scales of days or longer, other aspects, such as rectified currents, extend to the longest time scales. In addition, emphasis is often naturally placed on understanding the effects of different forcings one at a time. Ultimately, however, it will be necessary to deal with the combined effects of winds, waves, surface heat fluxes, tides, and buoyancy. This is an exciting frontier!

# 7

# Tides in the Coastal Ocean

## 7.1. Introduction: Continual Forcing and Responses

The study of tides is perhaps the oldest branch of physical oceanography. This, of course, makes sense for a number of practical reasons, such as the need to know sea level and currents while navigating into and out of a port. It is thus not surprising that tides were studied quantitatively as early as in the eighteenth century, by Laplace, and efforts continue to the present. For such a venerable line of research, it is still a very lively topic involving, for example, studying the implications of tides for ocean mixing, and hence global ocean circulation.

Tides in the coastal ocean are distinctly different from those in deeper water. Astronomical forcing on shelves, as in lakes, is not very effective, so tides over the continental margin are driven by neighboring oceanic conditions. An interplay of factors then determines the strength of tides on the shelf, with some coastal regions, such as Patagonia and northwest Australia, having dramatically large currents and sea level elevations. Coastal ocean tides, especially when they are strong, give rise to a range of consequences including mixing, biological productivity, and mean flow generation.

A typical tidal record, representing bottom pressure not far south of Martha's Vineyard (Figure 7.1), illustrates several points. The first thing to notice is the regularity: roughly (but definitely not exactly) twice every day (i.e., semidiurnally), there are clear cycles, but the amplitude of these cycles varies. Often, for example, on September 11, consecutive cycles alternate in amplitude: strong, weak, strong, weak. This difference is called the "'diurnal inequality," and it can be described as the interplay of two tidal frequencies. Further, there is a clear variation, with a period of roughly 2 weeks: the "spring/neap" cycle, when spring tides are the high-amplitude phase (around September 17), and neap tides are the low-amplitude times (around September 11 and 25). In addition, there are occasionally irregular events (like on September 7–8) that are not simply sinusoids but, rather, are associated with other, less regular, forcings, such as wind stress. The following sections explore some of the rationale for the regular, astronomically driven, cycles and then examine coastal tides and their consequences.

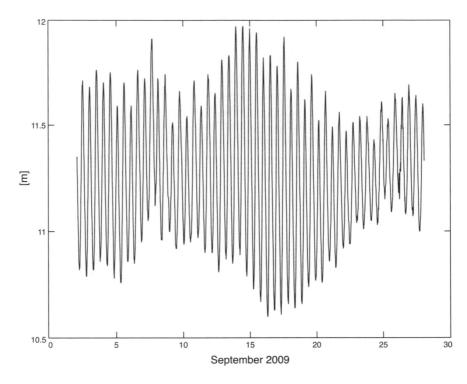

**FIGURE 7.1.** An example of observed tides in the coastal ocean, with bottom pressure (expressed as meters depth of water) at the Martha's Vineyard Coastal Observatory, in about 11 m of water, just south of Martha's Vineyard.

## 7.2.    Background: Forcing and Frequencies

At first glance, tidal frequencies appear to represent some rather odd periodicities. Indeed, it may be these periodicities that reputedly drove Aristotle to distraction when studying flow in a strait in Greece. This section summarizes the basic forcing for the tides and then rationalizes the multiplicity of frequencies, along with the general features of ocean-scale tides. The summary here is admittedly compact. Thorough treatments of the fundamentals can be found in Hendershott and Munk (1970), Cartwright (2000), Wunsch (2015), or Egbert and Ray (2017).

First, consider an idealized Moon-Earth orbiting pair (Figure 7.2). Both bodies revolve around the pair's center of gravity, which is relatively near Earth's center because of Earth's far greater mass. Now, consider forces acting along the line connecting Earth's and the Moon's individual centers of mass, where there are two forces at play. At Earth's surface on the side facing the Moon there is a net gravitational force $F_G$, which is dominated by Earth's gravity but also has a weaker upward contribution from the Moon's gravity. Second, there is the centrifugal "force" $F_C$ (computed relative to the center of mass of the Earth-Moon system), which reflects the tendency for a rotating body to move outward.

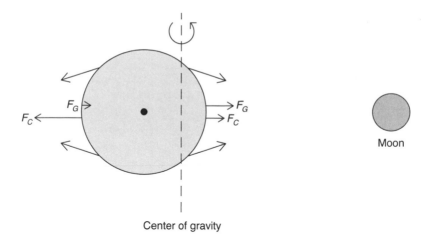

**FIGURE 7.2.** Schematic representation of tidal and centrifugal forces associated with the Earth-Moon system. $F_G$ is the tidal gravitational force (subtracting Earth's gravitation), and $F_C$ is the centrifugal force.

One can subtract from this sum of forces Earth's gravitational force, because it is very nearly the same everywhere on the surface. What is left (the Moon's gravitational force plus the centrifugal force) is the net tidal force. It is pointed directly upward along the axis of the center of masses, but for points on Earth's surface away from that axis, there is a substantial component of force parallel to the surface (Figure 7.2). Now, continuing on the Earth-Moon centerline to the other side of Earth, a body at the surface again feels gravity, with both Earth and the Moon now pulling downward. There is also an upward centrifugal force, which is stronger on this side of Earth because it is farther from the Earth-Moon center of gravity (axis of rotation). Subtracting the static effect of Earth's gravitation leads again to the tidal forcing. What makes all this particularly interesting is that the total upward (relative to Earth's surface) tidal forces along the Earth-Moon axis on both sides of Earth are identical, as a consequence of how, when considering all forces in the center-of-mass frame, the orbital forces acting on the entire bodies have already been accounted for. The basic result, of equal upward forcing bulges on *both* sides of Earth, does not depend on where the total center of gravity falls—whether within Earth, between Earth and the Moon, or within a distant body, such as the Sun relative to Earth.

Now, Earth takes a day to complete a rotation about its axis, so the Moon's location appears to observers on Earth to change through the night, but the relevant concern is the time it takes for the Moon to reappear at the same location in the sky. This means accounting not just for Earth's rotation relative to the "fixed" stars (i.e., with a period of a *sidereal day*: about 23 hours and 56 minutes) but also for the Moon's orbiting around Earth with a period of 27.32 days, so the Moon's position changes by 13.18° every day. Accounting for both of these effects leads to the *lunar day*—the time for the Moon to reappear at roughly the same point in the sky—of 24.84 hours. During the course of a lunar day the Moon passes overhead, and then the forcing bulge on the opposite side also

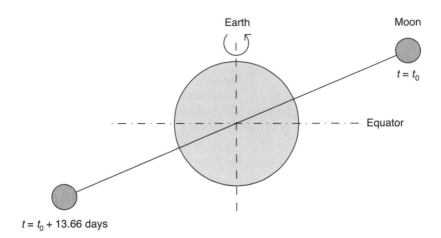

**FIGURE 7.3.** A schematic representation of the Moon's plane of orbit relative to Earth's equator. The angle between Earth's equator and the lunar orbital plane varies (over an 18.61-year cycle) between about 18.3° and 28.6°.

passes by. This means that at a fixed spot on Earth an upward tidal force is felt twice in a lunar day, or once every 12.42 hours. The lunar day thus defines a first fundamental tidal frequency, $f_1 = 1/(24.84 \text{ hours})$. Note that this and other fundamental tidal frequencies are given in cycles per unit time. Notice, too, that this frequency involves both Earth's rotation and the lunar orbital period.

The diagram in Figure 7.2 is, of course, too simple. Specifically, the Moon's orbital plane is not the same as that of Earth's equator (Figure 7.3). Nor is it aligned with the plane of Earth's orbit around the sun: the *ecliptic*. During the course of the Moon's 27.32-day orbit, the Moon is above Earth's northern hemisphere for half the period and spends an equal time above the southern hemisphere. This periodicity is important because the Moon's position structures the geometry of the tidal forcing on Earth. This hemispheric asymmetry is associated with another fundamental frequency (the inverse of a lunar month), $f_2 = 1/(27.32 \text{ days})$. This frequency is also sometimes called that of the lunar declination. Notice that one can recover the sidereal day using these two fundamental frequencies, that is, as $(f_1 + f_2)^{-1}$.

To this point only the lunar tidal forcing has been considered. The Sun actually exerts a much greater force on Earth, but its net tidal forcing (which accounts for centrifugal acceleration as well as gravitation) is only about half as great as the Moon's once centrifugal effects are accounted for. But the Sun cannot be ignored. Just as with the Moon, there are semidiurnal and diurnal components associated with Earth's rotation and with the tilt of its rotation axis relative to the ecliptic. Further, the period of Earth's revolution about the Sun matters, because Earth's axis is tilted about 23.4° relative to the ecliptic plane (Figure 7.4), so that during part of the year, the northern hemisphere tilts toward the Sun, and part of the year tilts away (thus accounting for Earth's seasons). This represents an effect similar to that of the tilt of the Moon's orbit relative to the equator, but with the repeat period of a year. Thus, there is a new key frequency, $f_3 = 1/(365.24 \text{ days})$.

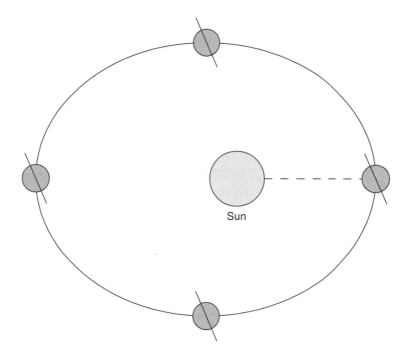

**FIGURE 7.4.** Schematic representation of Earth moving around the Sun over the course of a year. The tilted line running through Earth represents its north–south axis of rotation. The point is that the northern hemisphere tilts toward the Sun part of the year, and away from it part of the year.

Now, the frequencies $f_1$, $f_2$, and $f_3$ can be combined to obtain the frequencies associated with the other solar tides. For example, using logic similar to that for defining the lunar day, the solar semidiurnal tidal frequency is $1/(12 \text{ hours}) = 2f_1 + 2f_2 - 2f_3$. Thus, there is no need to introduce a new fundamental frequency that reflects, for example, the solar diurnal tide.

Other, lower frequencies also enter. One is $f_5 = 1/(18.61 \text{ years})$, associated with the tilt of the Moon's orbital plane relative to Earth's equator (Figure 7.3). The angle between these two planes oscillates between about 18.3° and 28.6°. This variation arises because none among the equatorial, ecliptic, and lunar orbital planes are coincident with another. A second very low frequency, $f_6 = 1/(20{,}940 \text{ years})$, is associated with the slow progression of Earth's elliptical orbit orientation relative to the Sun. Consider the dashed line in Figure 7.4 (representing the nearest point on the orbit to the Sun), and imagine it slowly rotating around the Sun. Frequency $f_4 = 1/(8.85 \text{ years})$ is the lunar equivalent of $f_6$.

Now, to this point it has been assumed that the tides themselves occur at the same frequencies as the forcing. This is certainly reasonable in a strictly linear system. However, linearity is not always a good assumption for tides. Consider the shallow-water equations (2.4.10, a form suitable for spatial scales small relative to Earth), which govern tides in an ocean with no density stratification. Either momentum advection (terms such as $\overline{u}\,\overline{v}_x$) or a quadratic bottom stress (3.6.7) can lead to generation of tidal harmonics, given

**TABLE 7.1.** Some Major Tidal Species

| Darwin notation | Doodson coefficients | | | Period | Proper name |
|---|---|---|---|---|---|
| | $[m_1$ | $m_2$ | $m_3]$ | | |
| $M_f$ | 0 | 2 | 0 | 13.661 days | Lunar fortnightly |
| $O_1$ | 1 | −1 | 0 | 25.819 hours | Basic lunar |
| $P_1$ | 1 | 1 | −2 | 24.066 hours | Basic solar |
| $K_1$ | 1 | 1 | 0 | 23.934 hours | Lunar-solar declinational |
| $N_2$ | 2 | −1 | 0 | 12.68 hours | Larger lunar elliptic semidiurnal |
| $M_2$ | 2 | 0 | 0 | 12.421 hours | Lunar semidiurnal |
| $S_2$ | 2 | 2 | −2 | 12.00 hours | Basic solar |
| $K_2$ | 2 | 2 | 0 | 11.97 hours | Lunisolar semidiurnal |

forcing at a single frequency. Thus, a lunar semidiurnal tide (12.42-hour period) can give rise (among other harmonics) to a 6.21-hour component as well. The nonlinear generation of tidal harmonics, of course, is most appreciable when tidal velocities are particularly strong, which in practice means these effects are usually most obvious in shallow water, such as within estuaries.

The set of fundamental frequencies can be added and subtracted to come up with a range of tidal periodicities. The resulting frequencies are expressed in terms of the Doodson coefficients $m_j$: integers which multiply the fundamental frequencies $f_j$ to describe the major tidal forcing frequencies. For example, the lunar semidiurnal frequency $1/(12.42 \text{ hours})$, $= 2f_1$ is described by Doodson coefficients $[m_1 \, m_2 \, m_3] = [2 \, 0 \, 0]$. In general, a given tidal frequency $f_T$ can be written as

$$f_T = \sum_{j=1}^{6} m_j f_j. \tag{7.2.1}$$

Table 7.1 includes some frequently used examples.

Another popular way to describe tidal forcing frequencies is the Darwin notation, which consists of a subscripted letter. The letter describes the type of forcing (M for lunar and S for solar, and K for combined, for example), and the subscript describes whether it is semidiurnal (2: tides occurring about twice per day), diurnal (1: tides occurring about once per day), fortnightly ($f$: tides occurring about every 2 weeks), and so forth. Thus, $M_2$ is the semidiurnal lunar tide having Doodson coefficients[1] $[2 \, 0 \, 0 \, 0 \, 0 \, 0]$, and its harmonic $M_4$ has numbers $[4 \, 0 \, 0 \, 0 \, 0 \, 0]$. Again, some commonly observed examples are presented in Table 7.1.

---

1 It is common practice to add 5 to each of $m_2$, $m_3$, etc., but *not* to $m_1$ and then subtract 5 to compensate in (7.2.1). Thus, the $M_2$ Doodson coefficients would become $[2 \, 5 \, 5 \, 5 \, 5 \, 5]$. This convention is used so that none of the numbers are negative.

Once lower frequencies (Doodson coefficients 4–6) are accounted for, it becomes clear that the frequencies in Table 7.1 are approximations. In reality, $M_2$ actually represents a set of tidal frequencies/forcings, all having a period of very nearly 12.42 hours and with differing $m_4$ to $m_6$. From a practical standpoint, this does not matter too much, given the relatively short observational records, except that $f_4$ often needs to be accounted for as a correction. It is this "fuzziness" in frequency that accounts for the fact that one rarely sees the $M_2$ tidal frequency given with more precision than 12.421 hours, for example.

Returning to the actual tidal forces, these can be quantified using straightforward (albeit complicated) calculations based on the equations governing orbital mechanics. These forcing functions need to be corrected at some point for how tides make Earth itself bulge slightly (by tens of centimeters) and for the self-gravitation associated with tidal ocean mass redistributions. The tidal forces are normally written in terms of a tidal potential $\zeta_E$ so that, for example, the linearized, frictionless, depth-averaged equations of motion for an unstratified ocean can be written in spherical coordinates as

$$u_t - fv = -\frac{g}{a \cos(\varphi)}(\zeta - \zeta_E)_\lambda, \tag{7.2.2a}$$

$$v_t + fu = -\frac{g}{a}(\zeta - \zeta_E)_\varphi, \tag{7.2.2b}$$

and the hydrostatic balance still applies in the vertical. The coordinates $(\varphi, \lambda)$ are latitude and longitude, respectively, and $a$ is Earth's radius. The (eastward, northward) horizontal velocity is given by $(u, v)$ and free-surface height by $\zeta$, $\rho_0$ is a constant density, $f$ is the Coriolis parameter, and subscripts $(\varphi, \lambda, t)$ represent partial differentiation. The form (7.2.2) emphasizes that the relevant aspect of tidal forcing is not the upward component (which is small compared with Earth's gravity) but, rather, the component parallel to Earth's surface at a given location (Figure 7.2: vectors occurring off the line connecting Earth's and the Moon's centers of gravity). The tidal potential (in the form given here) is also called the *equilibrium tide*, because (7.2.2) appears to allow the solution $\zeta = \zeta_E + \zeta_0(t)$ (where the constant of integration, $\zeta_0(t)$ is chosen so that the areal average of $\zeta$ remains constant to conserve volume) as a solution. In general, this is not a viable solution, however, because the ocean has a finite response time for $\zeta$ associated with how long it takes for a wave to cross a basin, and because this simple solution does not allow for blocking flow through coastal boundaries. The magnitude of $\zeta_E$ depends on the tidal species and on location, but it is typically less than 0.5 m.

The tidal potential is global in scale; it does not vary substantially over distances of a couple hundred kilometers or less. Such a local system experiences a nearly uniform tidal potential, so there is little force differential to excite water motions within a small basin. As a result, the actual surface deviation $\zeta$ is small, for example, a few centimeters for a Laurentian Great Lake. Such weak tides can readily be masked by more energetic fluctuations owing to, for example, wind forcing. In contrast, for an ocean basin with extent comparable to Earth's radius, the force differential across the basin is substantial, and the response time (roughly the basin's scale divided by Poincaré wave speed) is not

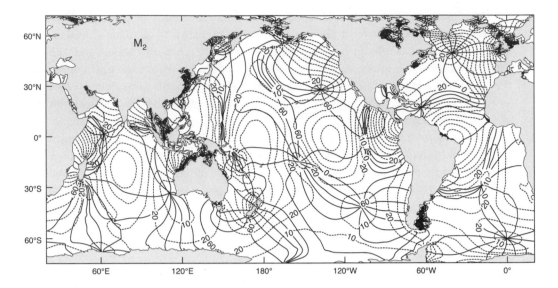

**FIGURE 7.5.** Global M$_2$ (12.42-hour) tidal elevation estimates using a combination of models and observations, including mainly satellite altimetry. Dashed contours are amplitude (10 cm contour interval) and solid contours are phase (60° contour interval). Adapted from Ray (1999).

rapid relative to a tidal period. Thus, tidal forcing is more effective and dynamic on the scale of an entire ocean. The result is that for a continental shelf (which is typically less than a couple hundred kilometers wide), direct astronomical forcing is usually not significant, while forcing by oceanic tides at the offshore boundary is important.

The global tidal problem is complicated by the geometry of the ocean basins: even at lowest order, the contorted boundaries and irregular bottom must be accounted for. A good deal of insight can be had by studying ocean basin modes (e.g., Platzman et al., 1981). These resonant barotropic wave structures are typically obtained using realistic, linearized numerical models of a world ocean. There are a number of modes with frequencies in the semidiurnal and diurnal tidal frequency range, and all have somewhat involved structures. The basin modes, however, lead to physical insight: often there are aspects that resemble a large-scale Kelvin wave (section 5.3) propagating around a basin or even something like a Rossby wave. To the extent that these modes occur near a tidal frequency and that their structure projects onto the structure of the tidal forcing, these modes are excited and can even dominate ocean-scale tides. Indeed, global tidal models (e.g., Figure 7.5) show structure and propagation that is readily interpreted in terms of Kelvin waves. For this M$_2$ sea level example, in the central North Atlantic, between Newfoundland and Britain, there is an *amphidrome*[2]: a minimum of amplitude where the entire range of phase lines come together. There is an amplitude maximum north of this

---

2 Appropriately, from the Greek word for "running around," just as the tide's phase propagates around the amphidrome.

site that extends from western Europe across to Canada and even (to a lesser extent) along the eastern United States. The phase lines associated with this feature correspond to westward propagation across the Atlantic and then southward along North America. This entire structure is associated with a Kelvin wave–like basin mode. Similar spiderlike phase features are found in all the major ocean basins.

Basin-scale tidal models have reached a very high degree of sophistication (e.g., Egbert and Ray, 2017) in that they assimilate satellite altimeter data and include ocean stratification. Aside from their remarkable predictive value, these models lead to other interesting insights. One is with regard to global tidal dissipation, for which there are two major causes. One (roughly a third of the total tidal dissipation) is the generation of internal waves ("internal tides") when tidal flow crosses bathymetric features such as the Hawaiian Ridge. These internal waves radiate away and eventually dissipate, thus contributing to ocean mixing. The second major loss (about two-thirds of the dissipation) occurs when tidal currents reach into shallow water, amplify, and dissipate via bottom friction. These mechanisms are discussed further in the following sections.

The stage is now set for treating tides over the continental shelf. Because tides are not generated substantially in the relatively narrow coastal ocean by astronomical forcing, shelf tides occur in response to conditions in the adjoining open ocean.

## 7.3.    Generation of Tides on the Shelf

Some simple, linear barotropic models provide a good deal of insight on tides over the continental shelf. The approach is to ask how shelf/slope waters respond to oceanic tides when they are expressed as periodic signals imposed where the continental slope abuts the relatively flat deep ocean. It is then possible to solve for tides over the shelf given

$$u_t - fv = -g\zeta_x - \frac{\sigma_F}{h}u, \tag{7.3.1a}$$

$$v_t + fu = -g\zeta_y - \frac{\sigma_F}{h}v, \tag{7.3.1b}$$

$$0 = \zeta_t + (hu)_x + (hv)_y, \tag{7.3.1c}$$

where $(x, y)$ are the cross-shelf and alongshore coordinates, $(u, v)$ are the corresponding velocity components, $\zeta$ is the free-surface elevation, $g$ is the acceleration due to gravity, and $\sigma_F$ is a bottom resistance parameter. The coast is at $x = 0$, and the water depth $h$ is taken to depend primarily on $x$. The use of Cartesian coordinates and neglect of the tidal potential are justified because continental shelves are generally narrow relative to Earth's radius.

Assume that the tides at the shoreward edge of the adjoining deep ocean have the form

$$\zeta = \zeta_0 \exp[i(ly - \omega t)] \qquad \text{at } x = L, \tag{7.3.2}$$

where the bottom is flat for $x > L$. Thus, the solution over the continental margin, $0 < x < L$, similarly has the form

$$\zeta = Z(x) \exp[i(ly - \omega t)]. \tag{7.3.3}$$

For purposes of illustration, the frictional terms in (7.3.1) are now deleted. This is a good approximation when

$$\frac{\sigma_F}{\omega h} \ll 1, \tag{7.3.4}$$

a condition which is not always satisfied in reality. It is then straightforward to use (7.3.1) and (7.3.3) to obtain a barotropic vorticity equation:

$$0 = Z_{xx} + \frac{h_x}{h} Z_x + \left( \frac{\omega^2 - f^2}{gh} - l^2 - \frac{flh_x}{\omega h} \right) Z. \tag{7.3.5}$$

To understand the system better, it is helpful to use a highly idealized, constant-slope topography

$$h = \alpha x \qquad \text{for } x < L. \tag{7.3.6}$$

Further, following Clarke and Battisti (1981), consider the alongshore scale, $l^{-1}$, of the tide. Physically, this can represent spatial variability in the open-ocean tide, which generally has scales of $O(1000 \text{ km})$ (e.g., Figure 7.5). Alternatively, alongshore variability can also be introduced by topography: bays, peninsulas, or variations in shelf width. For now, assume that the alongshore scales are large in the sense of $lL \ll 1$; that is, the topography is reasonably smooth. Using topography (7.3.6) and the smoothness assumption, (7.3.5) becomes

$$0 = Z_{xx} + \frac{1}{x} Z_x + \frac{\mu}{x} Z, \tag{7.3.7a}$$

where

$$\mu = \frac{\omega^2 - f^2}{g\alpha} - \frac{fl}{\omega}. \tag{7.3.7b}$$

Solutions to (7.3.7) for $\mu > 0$ are Bessel functions (e.g., Abramowitz and Stegun, 1965) of $\xi = 2 (\mu x)^{1/2}$; that is, they are cross-shelf standing Poincaré waves (section 5.3), modified by the nonuniform depth. Thus, a superinertial, $\omega > |f|$, resonance is possible, most likely of one-quarter wavelength, and strong shelf tides are quite possible, since semidiurnal tides are superinertial nearly everywhere on Earth. In addition, depending on the frequency relative to $f$, resonances with shelf waves, Kelvin waves, or edge waves are possible (Figure 5.5), but of these other types, the shelf wave seems most likely to produce substantial currents and then only for diurnal tides (e.g., section 5.9).

There are in fact many spectacular examples globally of quarter-wave shelf tidal amplification. A well-known case is the Gulf of Maine–Bay of Fundy system off eastern North

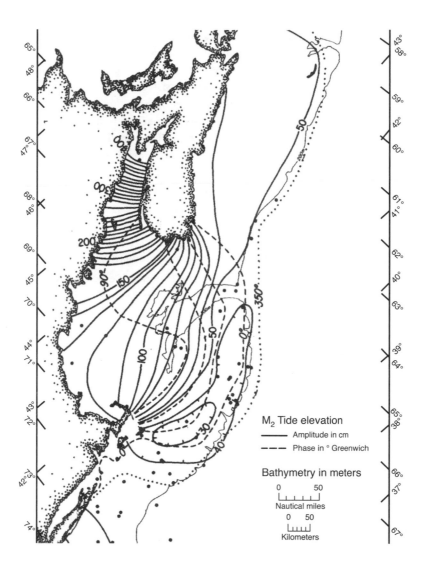

**FIGURE 7.6.** An example of tidal resonance over the shelf. Surface elevation amplitude (solid contours, in cm) and phase (dashed contours) for the $M_2$ tide in the Bay of Fundy–Gulf of Maine system. The 200 m isobath is also shown as a thin, solid contour. Adapted from Moody et al. (1984).

America, where the $M_2$ sea surface amplitude exceeds 4 m near the head (Figure 7.6). Garrett (1972) used tide gauge data there to estimate a resonant period of about 13.3 hours and a resonance quality $Q_T$ of around 5 ($Q_T$ is a measure of resonance "sharpness"; it is larger for strong, weakly damped peaks). This $Q_T$ value indicates that although the time derivative terms in (7.3.1) generally dominate frictional terms, dissipation is certainly important in constraining the resonant response here. This conclusion is, of course, not surprising given that global tidal models show that frictional dissipation of strong tides in shallow regions plays the strongest role in global tidal dissipation.

The phase structure in Figure 7.6 is consistent with this dissipation. Along a straight shelf with no dissipation and real $l$, solutions to equations (7.3.7) resemble standing waves in the cross-shelf direction: for a given $y$ everything is either in phase or 180° out of phase as a function of $x$ over the shelf. This is consistent, at superinertial frequencies, with equally and oppositely propagating onshore/offshore Poincaré waves. Including friction in the problem induces cross-shelf phase changes consistent with the onshore-propagating wave having a greater amplitude than the reflected offshore-propagating wave that has undergone frictional decay in shallow water. This difference means that there is net energy radiation onto the shelf, where it is at least partially dissipated. This net onshore propagation is most obvious in Figure 7.6, where frictional effects are strong over the shallow, $O(50 \text{ m})$ deep, Georges Bank, which is located near the ocean end of the embayment.

Often, in the nonresonant case, $\mu^{-1}$ is fairly large—in the range of 300–1000 km. When $\mu L \ll 1$, some useful and insightful simplifications are possible (Clarke and Battisti, 1981). Specifically, a power series expansion of the Bessel function is efficient, or, even simpler, the solution to (7.3.5) can be found directly as a series expansion. To proceed, Clarke and Battisti took the forcing and response to be proportional to $\exp(-i\omega t)$, and they further assumed (and justified) it is reasonable to account for alongshore tidal variations as

$$il = Z_y / Z, \tag{7.3.8}$$

where $l$ does not vary across the shelf and where $|lL| \ll 1$, a reasonable assumption for smooth topography given the oceanic scale of offshore tides. The inverse scale $l$ is allowed to be complex to account for alongshore variations in amplitude, as well as phase. Their analysis can also allow for frictional dissipation. Further, to obtain a more realistic topography, they described the actual shelf-slope profile in terms of a sequence of straight-line fragments of the form

$$h = a\,(x - b) \tag{7.3.9}$$

that are matched up to obtain a continuous, but angular, depth profile. As long as $\mu$ is small, simple low-order polynomial solutions to

$$0 = Z_{xx} + \frac{1}{(x-b)} Z_x + \frac{\mu}{(x-b)} Z \tag{7.3.10}$$

(from 7.3.5) are then found in terms of $\xi = 2[\mu(x-b)]^{1/2}$ (each segment having its own local $\alpha$, $\mu$ and $b$) and matched to obtain a continuous solution for pressure. Their results usually describe a general increase in the $\zeta$ amplitude as the coastline is approached from offshore. Applying their simple model for a range of locations, Clarke and Battisti (1981) found it works well, given information about tides offshore (including $l$), when the model assumptions, such as slow alongshore variations in $\zeta$ or $h$, are valid. They realistically reproduced strongly enhanced shelf tides for semidiurnal forcing (i.e., where resonant Poincaré waves are possible), for example, off northern Australia and off Argentina. Arbic et al. (2009) since showed that when shelf tides are greatly enhanced, the deep-ocean tides are, in turn, affected by the shelf tides, so it is no longer reasonable to treat the shelf as responding to a strictly prescribed ocean. That is to say that boundary condition (7.3.2) is not always a reasonable approximation even though it is illustrative.

## 7.4.    Tidal Rectification

Oscillating tidal currents can transport momentum across isobaths and thus generate mean flows. The basic physics is sketched out here, but the problem was treated much more carefully and generally by Huthnance (1973) and by Loder (1980). Consider an unbounded region with a sloping (in the $x$ direction) bottom. Further, there is an imposed cross-isobath tidal current $u_0(x) \sin(\omega t)$, and (by assuming a rigid-lid approximation locally) the cross-isobath transport amplitude $U_0$ is constant, so that $u_0 = [U_0 / h(x)]$, where $U_0$ is driven at a reference location where the depth is large compared with the local depth $h(x)$ of interest. Assuming no along-isobath variations, and no density variations, the depth-averaged along-isobath momentum equation is then

$$v_t + uv_x + fu = -\sigma_F v/h. \tag{7.4.1}$$

Neglecting for now the nonlinear term, it is then straightforward to use $u = u_0(x) \sin(\omega t)$ and solve for the tidal along-isobath current

$$v \cong \frac{fU_0}{\omega h} \cos\left(\omega t + \frac{\sigma_F}{\omega h}\right), \tag{7.4.2}$$

where it has been assumed that $[\sigma_F/(\omega h)]^2 \ll 1$[3]. Now, the time average of (7.4.1) is, to the same degree of approximation,

$$\langle uv_x \rangle = -\sigma_F \langle v \rangle / h, \tag{7.4.3}$$

where the angle brackets denote a time average. It is then straightforward to use the expressions for $u_0$ and tidal $v$ (7.4.2) to obtain

$$\langle v \rangle \cong -\frac{fh_x}{\omega^2 h} u_0^2 = -\frac{fh_x}{\omega^2 h^3} U_0^2. \tag{7.4.4}$$

This Eulerian mean along-isobath flow is in the same direction as long topographic wave propagation. An interesting property of this solution is its independence (at this order) of bottom friction, even though the friction plays a critical role by taking $u$ and $v$ out of quadrature with each other, thus allowing a cross-isobath eddy momentum flux. All told, both the forcing (left-hand side in eqn. 7.4.3), and the dissipation (right-hand side) that limits the acceleration, are proportional to $\sigma_F$.

The results up to this point have been highly idealized, so it is worth noting a few further aspects. First, since the rectification process involves nonlinear coupling in (7.4.1),

---

3 Huthnance and Loder both performed more rigorous derivations that do not require this small-friction assumption. Their expressions for mean Eulerian flow thus differ somewhat from (7.4.4), but the general form is similar.

a complete solution will involve higher tidal harmonics, as well as other (sum and difference) frequencies when more than one tidal species is present (e.g., Münchow et al., 1992). Next, the mean velocity (7.4.4) is strongly dependent on the water depth $h$, so rectification is most effective in shallow water and where the bottom slopes strongly. The vertical structure changes somewhat when the water column is stratified (Maas and Zimmerman, 1989) in that the mean flow is bottom intensified, but the result remains dependent on total water depth, because the tidal strength, $u_0$, still depends on $h$. Further, this sort of rectification becomes weaker but still operates when the isobaths are no longer straight (i.e., when the alongshore length scale is comparable to, or smaller than, $fu_0/\omega^2$; Brink, 2010), but a good deal more could be done on this topic. Fourth, Loder (1980) computed the along-isobath tidal Stokes drift[4] and showed it is not small compared with $<v>$ (7.4.4) and thus contributes importantly to the Lagrangian mean velocity. Finally, observing tidal rectification is sometimes complicated by other tidal effects (see next section) so it can be hard to isolate in practice.

## 7.5.   Tidal Mixing and Fronts

Strong tidal currents can sometimes generate enough near-bottom turbulence to mix well-stratified waters completely. To quantify the mechanism, it is useful to consider first the energetic implications of mixing and stratification. Say there is a linearly stratified water column (Figure 7.7, left panel) with its center of gravity located below the mid-depth of the water column. This happens because the deeper water is denser, so there is more mass below mid-depth than above. Now, if the water column is completely mixed (right panel), the center of gravity is at exactly mid-depth, because the density is now uniform. The point is that mixing leads to an elevation of the center of gravity of any stably stratified system, and thus, the gravitational potential energy of the system increases during homogenization.

The relationship between potential energy and mixing can be exploited to understand where surface-to-bottom mixing might be expected in the coastal ocean (Simpson and Hunter, 1974) if it is assumed that only one-dimensional (vertical) processes are effective; horizontal mixing and advection are neglected. The depth-integrated gravitational potential energy is

$$\Phi = \int_{-h}^{0} (\rho - \check{\rho}) gz \, dz, \tag{7.5.1}$$

where $\check{\rho}$ is a constant reference density. Then, the equation describing changes in potential energy $\Phi$ becomes

$$\frac{d\Phi}{dt} = -\frac{\gamma g F_T h}{2C_P} + \varepsilon_B c_{DB} \rho_0 \langle |u_B|^3 \rangle + \varepsilon_W c_{DA} \rho_A \langle |W|^3 \rangle, \tag{7.5.2}$$

---

4 Stokes drift is the net motion of particles in a purely oscillatory flow. In this context, it arises because $u$ and $v$ are larger in shallower water: when a water parcel completes a period of oscillation, it does not come back to where it started. See section 6.3 for a discussion of this effect in the context of surface gravity waves.

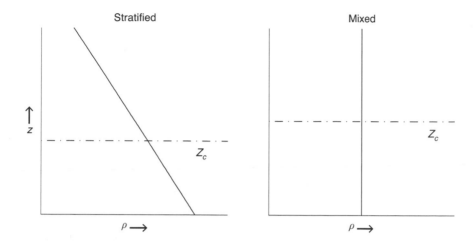

**FIGURE 7.7.** Schematic density profiles showing how the center of gravity $z_c$ is lower in a stably stratified water column (left) than in a well-mixed water column (right).

an expression comparable to (3.5.11), except that it also accounts for turbulence due to bottom friction. The first term on the right-hand side describes the decrease of potential energy due to warming at the surface, which makes the water there lighter and more stably stratified ($F_T$ is a surface heating rate, $\gamma$ is a thermal expansion coefficient, $h$ is the water depth, and $C_P$ is the specific heat of water). The remaining terms quantify the tendency to destroy stratification. The second term describes the effect of mixing created by turbulence generated at the bottom boundary ($\varepsilon_B$ is an efficiency factor, $c_{DB}$ is the bottom drag coefficient—see section 3.6, and $u_B$ is a representative velocity near the bottom). Finally, the last term represents the effect of wind-driven mixing ($\varepsilon_W$ is an efficiency factor, $c_{DA}$ is a surface atmospheric drag coefficient, $\rho_A$ is the density of air, and $W$ is a near-surface wind velocity). Whether mixing or stratification wins out is a matter of which terms are bigger on the right-hand side of (7.5.2), and $d\Phi/dt$ is exactly zero at the boundary between mixing and stratifying waters ($\Phi$ is in a steady state at this boundary).

The mixing terms in (7.5.2) are, of course, always positive, and the warming term is typically only negative during summertime warming (when $F_T$ is positive). Shallow waters should be completely mixed during the winter ($F_T < 0$) if there is no horizontal buoyancy transport to stabilize the system. During summer the right-hand side can be positive if mixing is strong enough to overcome surface warming. In some places in the coastal ocean, such as near the British Isles or over Georges Bank, near-resonant tidal conditions lead to strong currents, $O(0.5\text{–}1 \text{ m/s})$. Because water is so much denser than air, this means that the tidal bottom stress is about equivalent to that of an atmospheric hurricane, and so in these locations, tides dominate over the typical winds in terms of mixing impact. Thus, in regions of strong tides, the critical condition occurs when the warming and bottom stress terms in (7.5.2) approximately balance, that is, when

$$\frac{h}{\langle |u_B|^3 \rangle} \cong \frac{2\varepsilon_B c_{DB} C_P \rho_0}{\alpha g F_T}. \tag{7.5.3}$$

On a contour map of $\dfrac{h}{\langle |u_B|^3 \rangle}$, the value (7.5.3) should separate stratified water (where warming dominates) from well-mixed water (where tidal mixing dominates). Owing to continuity considerations, tidal currents very often strengthen in shallower water, so $u_B$ usually depends strongly on $h$. Thus, $\dfrac{h}{\langle |u_B|^3 \rangle}$ contours often, at least locally, parallel isobaths, and so the frontal boundary between mixed and stratified waters also tends to follow depth contours. Actual frontal locations, of course, vary somewhat temporally, because tidal currents vary over time scales of hours to weeks and because surface heating also varies on diurnal, weather, and longer time scales. Alternative formulations exist (e.g., Loder and Greenberg, 1986) for the location of a tidal mixing front, but they are difficult to distinguish observationally or numerically from the Simpson/Hunter criterion, since they all depend on $h$ divided by some function of $|u_B|$.

Numerical tidal models can be used to estimate fluctuating currents, hence $\dfrac{h}{\langle |u_B|^3 \rangle}$, so that results can be compared with observed tidal mixing fronts (e.g., Pingree and Griffiths, 1978, Timko et al., 2019). The comparisons are strikingly good in a wide range of settings with tidal mixing fronts: the Yellow Sea, the Bering Sea, the Argentine shelf, and in Hudson Bay, to name just a few examples.

One major simplification in the Simpson/Hunter analysis is the assumption of one-dimensionality—that lateral advection is not important at lowest order. That the Timko et al. (2019) model, which includes wind forcing and is fully three-dimensional, often makes predictions comparable to the Simpson/Hunter theory argues that, in many cases, the surface heat flux is indeed the dominant stabilizing effect. However, this is clearly not true in some cases. Exceptions can occur when horizontally advected water parcels of uncertain origin temporarily stabilize otherwise mixed waters (e.g., on Georges Bank: Dale et al., 2003). A particularly compelling and relatively sustained example, however, is that of relatively fresh water flow out of an estuary into waters that would otherwise be tidally mixed. The affected areas, called ROFIs (regions of freshwater influence), can give rise to episodically stable stratification on time scales of hours to fortnightly (i.e., spring/neap tidal variations) or longer (e.g., Simpson, 1997). In this case, the tidal mixing, in turn, can strongly affect the behavior of the buoyancy current.

The tidal mixing front's structure (e.g., Figure 7.8) has interesting implications, even in the absence of lateral advection. The front's density field suggests, through thermal wind balance, that there is an along-frontal jet. Indeed, Chen et al. (1995) used Georges Bank numerical models and ocean data to demonstrate that these nearly geostrophic currents exist and that the frontal current varies seasonally as the surface heat flux (hence frontal location, structure, and density contrast) varies. In fact, it appears that their frontal jet, which is in the same direction as the tidally rectified flow (section 7.4), is stronger than the tidal rectification, at least during summer. More extensive observations (Brink et al., 2003) show that the along-isobath flow on the northern side of Georges Bank is indeed much stronger and better defined in the summer (about 0.2 m/s mean) than the winter (near zero mean), a seasonal variation that would not occur if only barotropic tidal rectification (section 7.4) were operating. It appears that the strong depth sensitivity evident in (7.4.4) makes the barotropic rectification relatively weak at this 50 m deep

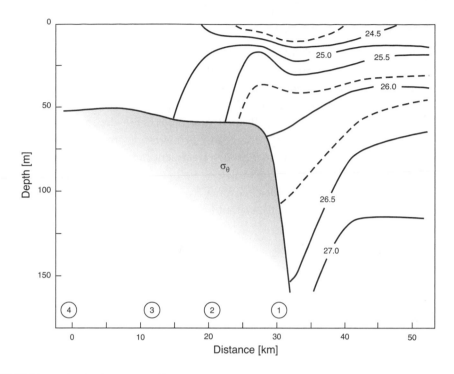

**FIGURE 7.8.** A tidal mixing front: a typical density ($\sigma_t$) section across the northern side of Georges Bank, July 3, 1988. Adapted from Loder et al. (1992).

location. Chen et al. (1995) also modeled the circulation in the cross-frontal plane and found it to be rather complex and to have an important Stokes drift component, so that a purely Eulerian description is clearly incomplete.

Homogenized regions bounded by tidal mixing fronts, such as over Georges Bank or around the British Isles, are often associated with high biological activity (e.g., Simpson and Sharples, 2012). Numerical model experiments clarify a key mechanism involved (Franks and Chen, 1996). Specifically, for a brief (a couple hours) period during each tidal cycle, ambient stratified waters (with higher nutrient concentrations at depth) extend into shallower areas at a time when tidally driven vertical mixing is particularly strong. The result is an upward nutrient flux concentrated near the shallow-water (well-mixed) side of the front. Thus, there is a regularly repeating injection of new nutrients (meaning nutrients that do not stem from local recycling) into the tidally mixed region, and these lead, typically, to high phytoplankton concentrations near the front.

## 7.6.   Internal Tides

When a barotropic tide encounters a sloping bottom, the cross-isobath flow $u$ must be balanced by a vertical velocity $w$, as expressed by the bottom boundary condition

$$0 = w + h_x u \qquad \text{at } z = -h(x), \qquad\qquad (7.6.1)$$

where $x$ is directed across isobaths, and $h(x)$ is the water depth. In an unstratified ocean, this bottom boundary condition is accounted for in depth integrating (or vertically averaging), so it is not very obvious. But in a stratified ocean, this induced vertical motion displaces density surfaces vertically and thus excites internal waves (assuming the tidal frequency is greater than $f$). These tidally excited, strongly depth-dependent currents are called the *internal tide*. The topic embraces a range of interesting processes, some of which are briefly summarized here. A more thorough treatment can be found, for example, in Vlasenko et al. (2005).

In a resting, stratified ocean with a buoyancy frequency $N$, internal waves can propagate both horizontally and vertically, so that for a constant $N$, the wave's frequency $\omega$ obeys

$$\omega^2 = f^2 \sin^2\theta + N^2 \cos^2\theta, \qquad\qquad (7.6.2)$$

where $\theta$ is the angle between the horizontal plane and the direction of the wavenumber (and phase velocity) vector. (For a good introduction to internal wave propagation, see Gill, 1982.) The internal waves are resonantly excited when the wave's orbital velocity is parallel to the bottom. This requires the wavenumber vector (thus phase velocity) to be orthogonal to the bottom. Because the group velocity is orthogonal to the phase velocity, energy propagation is then parallel to the bottom where the resonance occurs. The net effect is an internal wave beam propagating upslope and/or downslope parallel to the bottom location where the wave vector's slope equals $h_x$. Figure 7.9 illustrates this point using results from a linear internal tidal model having $N^2(z)$ from Figure 5.3. Note that near radius 42 km, $z = -3000$ m, the ridge of high variance (the wave beam) is parallel to the bottom. Once the beam intersects the bottom elsewhere (near radius = 38 and 46 km), it reflects and carries internal wave energy upward or offshore, respectively. Depending on geometry, it is possible for reflections to allow wave energy to pass into shallower water as well. The complexity of the beam patterns in the upper 1000 m of the water column in Figure 7.9 reflects the complexity of the realistic $N^2$ (Figure 5.3) in this model.

The exact path of an internal wave beam depends sensitively on the stratification and on ambient currents (e.g., Mooers, 1975). In the real coastal ocean, this means that the detailed geometry of a beam varies from time to time as currents or stratification change. Thus, for observations at a fixed location, the internal tide can be very intermittent, depending on whether a beam passes that point or not. Presumably, the closer the observer is to the beam's generation point, the less intermittent the tides will be, given that there is less distance over which perturbations can act to randomize the path. Thus, it is no surprise that internal tides in the coastal ocean are observed to be very intermittent, (e.g., Rosenfeld, 1990). To this point only linear internal tides, in which the internal waves occur at the same tidal frequency as the forcing, have been discussed. Linearity is not always a good approximation, however.

The next step is to explore a mechanism for generating nonlinear internal waves. Very often, when the bottom is flat, solutions to linear wave problems with stratification are

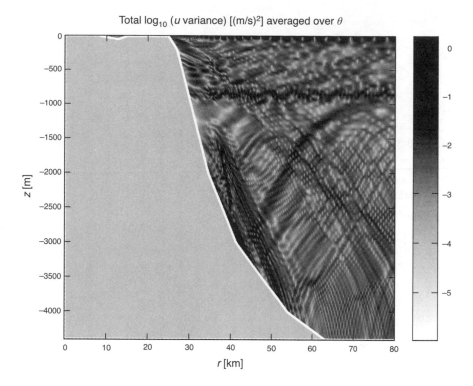

Total $\log_{10}$ ($u$ variance) [(m/s)$^2$] averaged over $\theta$

**FIGURE 7.9.** Modeled internal tidal paths. Shown is a cross section of the logarithm of azimuthally averaged variance of cross-isobath currents (m$^2$/s$^2$) for a linear model of a circular island in a stratified ocean. The forcing is a spatially uniform oscillating current at the M$_2$ tidal frequency (12.42-hour period). The radius (distance from island center) is $r$. The depth profile roughly mimics that around Bermuda, and the realistic $N^2$ profile is from near that island (Figure 5.3), but $f = 2 \times 10^{-5}$ 1/sec. Calculated as described in Brink (2021).

expressed in terms of a sum of baroclinic modes (see section 5.4). Each mode can be thought of as representing a balanced pair of phase-locked upward- and downward-propagating internal waves reflecting off the flat surface and bottom. Even complex wave beam patterns, such as those in Figure 7.9, can be expressed as a sum over many baroclinic modes so long as the bottom is flat and there are no ambient currents. One advantage of the modal approach is that each mode has a phase speed $c_n$ associated with it, which represents the speed at which a long (compared with water depth) wave would propagate in the absence of rotation. The wave speed decreases as the vertical modal structure becomes more complex (more zero crossings) with increasing $n$ (Figure 5.3). For a given $n$, $c_n$ increases with stratification, $N^2$, and with overall water depth. A simplification of this concept is to treat the ocean as having two layers, each with its own constant density, an idealization that allows a barotropic mode and only a single internal mode.

Knowledge of modal structures clarifies an important aspect of internal waves. Whereas in the deep ocean, internal waves can travel quickly relative to current speeds (e.g., $c_1$ can be greater than 2.5 m/s at midlatitudes), in shallower coastal waters, $c_1$ is often more like 0.5 m/s or less in the summertime and even less in the winter, when waters are weakly stratified. This basic phase speed $c_1$ can be thought of as a crude estimate of the maximum speed with which a linear internal wave can propagate. Hence, in the coastal ocean, internal wave propagation can easily be slow enough for the wave propagation speeds to be comparable to coastal currents, especially those associated with strong tides. This means that internal wave propagation can be accelerated, decelerated, or even reversed by ambient flows. Further, the orbital velocities associated with these waves can be of the same order as the linear propagation speed.

Thus, the internal tides in the coastal ocean are sometimes quite nonlinear in character, forming bores, solitons (a class of single-peak wave whose dynamics express an enduring balance of dispersive spreading and nonlinear sharpening), or trains of solitary waves (e.g., Helfrich and Melville, 2006) that can potentially travel coherently over large distances. Well-resolved measurements from the northern flank of Georges Bank (Figure 7.10) illustrate how tides can directly drive these waves. When the tide flows northward, a hydraulic jump forms just beyond the northern edge of the bank (Figure 7.10, first panel near the letter M, or see the upper panel of the schematic Figure 7.11). That is to say that locally the current is strong enough to prevent internal wave propagation upstream. (See Pratt and Whitehead, 2007, for a thorough treatment of hydraulic jumps, which occur where the ambient flow balances the opposing wave speed and which represent abrupt changes in layer thickness.) Specifically, there is a steep downward bend of the isopycnals, and then, proceeding northward, the nearly stationary jump occurs. South of the jump, internal wave propagation upstream (toward the bank) is impossible, because the opposing tidal current is faster than internal wave propagation in this shallow water. North of the jump, internal waves can propagate freely in either direction, since the tidal current is weaker (by continuity) and wave propagation faster (because of deeper, more stratified water). Once the tide reverses toward the south, internal wave radiation in either direction is possible north of the bank, including propagation onto the bank (i.e., now in the same direction as the current). Thus, the 20 m pycnocline deflection in the first panel is free to radiate away in either direction (Figure 7.10, third panel, features M and N; lower panel of Figure 7.11). Given the short horizontal scale and sometimes extreme magnitude of the jump, it can form into various sorts of nonlinear internal waves as it radiates away. The point of this example is that in the coastal ocean, the combination of topographic gradients, strong tides, and relatively shallow water provides a breeding ground for very nonlinear internal waves, which can propagate either toward shallow water or away, depending on tidal phase. On a given continental shelf, nonlinear internal waves can be generated at a local bottom irregularity or shelf edge, or they may propagate onto the shelf from distant generation sites (e.g., Nash et al, 2012). The properties of the consequent radiating internal waves are set by nonlinear interaction with the flow field, stratification, and bottom topography.

Once generated at a strong topographic feature, the nonlinear internal waves can then propagate onto the shelf. Sometimes, it is reasonable to treat the waves as solitons; in other cases, the waves might be more nearly like bores (an advancing near-discontinuity) or trains of isolated waves (e.g., Helfrich and Melville, 2006). Collectively, these nonlinear waves are generally called "solibores." In any case, the nonlinear

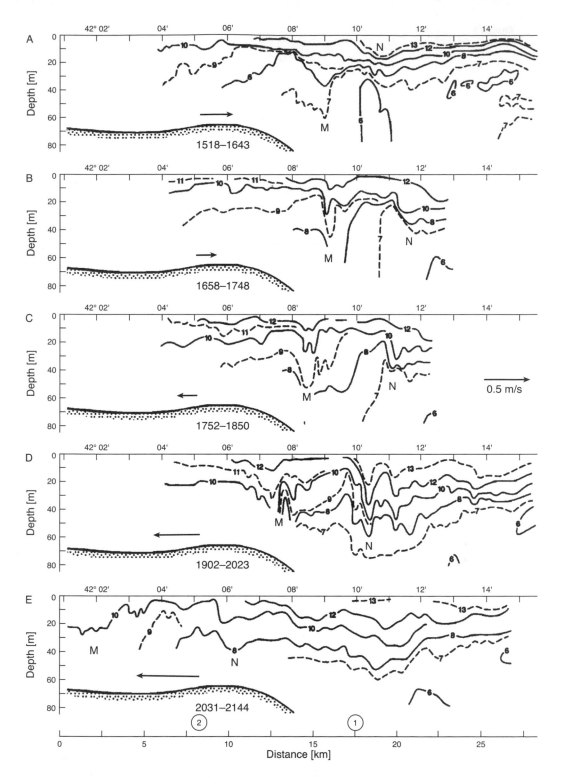

**FIGURE 7.10.** Observations of internal tidal generation. A sequence of well-resolved temperature sections across the northern flank of Georges Bank, with the sampling time range (UTC: e.g., 2031–2144 = 8:31–9:44 p.m. on the lowest panel), given in each panel. Horizontal arrows indicate tidal currents. North is toward the right. Adapted from Loder et al. (1992).

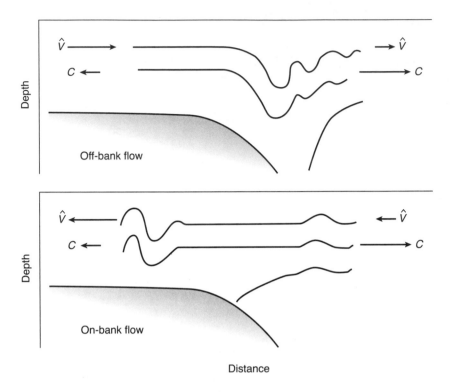

**FIGURE 7.11.** A schematic representing the processes in Figure 7.10. Upper panel: strong northward tidal flow. Lower panel: southward flow. $\hat{V}$ is the tidal velocity, and $c$ is the internal wave speed. Adapted from Loder et al. (1992).

internal waves typically occur in temporal groups at regular intervals, since they are often excited at only a particular phase of the tidal cycle (e.g., Shroyer et al., 2010). The strength or even presence of these groups is rather variable as well, since it depends on the variations in stratification, ambient currents, and spring/neap modulations in the barotropic tide's amplitude.

Internal solitons have the interesting property that the wave projects into the thicker of two layers; for example, if the lower layer is deeper, the interface disturbance is downward. As a soliton propagates onshore, the lower layer becomes shallow as water depth decreases, so that the wave changes sign, extending into the upper layer, as the coast is approached (e.g., Lynch et al., 2004). Although these waves can propagate into the inner shelf, some reflection has been observed as well (Davis et al., 2020). Remotely sensed information demonstrates that, in many cases, strong internal waves are generated at isolated locations, so the waves propagate in radially spreading patterns (e.g., Fu and Holt, 1982). It is thus not surprising that even on the inner shelf, internal wave conditions vary substantially over alongshore scales as short as 10 km (McSweeney et al., 2020). As the waves propagate into shallow water they can at least temporarily transport deeper, nutrient-rich waters onshore and thus potentially affect biological processes in kelp forests or on reefs (e.g., McPhee-Shaw et al., 2007; Davis et al., 2020).

As the tidally driven internal waves propagate onshore they experience a good deal of turbulent dissipation and can thus contribute substantially to overall water column mixing (e.g., Pritchard and Weller, 2005). These waves can generate turbulence by a number of mechanisms (e.g., Lamb, 2014). Perhaps the most obvious is associated with the wave's bottom friction. This can be supplemented by enhanced turbulence associated with the beam generation that occurs when the ambient tidal frequency and stratification (7.6.2) allow the linear internal tide phase propagation to be orthogonal to the local bottom slope ("critical reflection," as discussed previously). Nonlinear internal waves can also lose energy by radiating into the local internal wave continuum through scattering (Helfrich and Melville, 2006). Intense shears at the interface of upper and lower layers can generate turbulence, and in shallow enough water, the internal waves can steepen enough to break, just as surface waves do in the surf zone (e.g., Helfrich and Melville, 2006; Lamb, 2014; Becherer et al., 2021). Thus, the enhanced mixing associated with onshore internal wave propagation can be distributed across much of the shelf.

## 7.7.    Topographic Effects

One important aspect of tides is their dissipation in shallow waters. Bottom friction and water column mixing can be quantified in models and observations, but these are not the only important forms of dissipation. Irregular topography and the consequent topographic drag have been ignored so far. This drag arises, for example, when pressure on one side of a ridge or cape is greater than on the other, so a net force couples the topography with the ocean. The alongshore component of this topographic drag (also called form stress) due to bottom topography can be expressed as

$$D^y = \int_{-\infty}^{\infty} p_B h_y dy, \tag{7.7.1}$$

where $h$ is the water depth, $y$ is the alongshore coordinate, and $p_B$ is the bottom pressure. There is a similar expression for force on coastline irregularities as well. This sort of drag can act on scales from about a kilometer or less up to (in the atmosphere) global. The integral (7.7.1) is zero if both pressure and bottom topography are symmetric, say, around $y = 0$. Thus, for there to be a drag on the water column, there needs to be something about the flow that makes conditions upstream of a bump different from those downstream.

There are at least two mechanisms for creating this asymmetry. One is the introduction of a lee eddy or a hydraulic jump by flow around a cape or over topography: these strongly nonlinear features form downstream of an irregularity and occur for particularly strong flows. The second, perhaps more familiar, form of asymmetry is a lee wave. On scales that are often found in a coastal context, this would normally be an internal wave, but on larger scales, lee coastal-trapped waves or lee Rossby waves appear possible. A steady lee wave forms when a wave's upstream phase velocity is balanced by the downstream ambient flow, creating a steady-flow feature (see Gill, 1982, for an excellent discussion). However, with dispersive waves, the group velocity is not equal to the phase velocity, so energy is transported off to dissipate elsewhere, even though the phase velocity is nullified by the ambient current. With a fluctuating flow, similar physics can apply, although the mathematics becomes more complicated (e.g., Bell, 1975).

Do we know that this form drag (7.7.1) is an important effect in reality? Observations of form stress in the ocean are extremely rare because of the difficulty of knowing the exact depth (to within centimeters) of the water column where a pressure gauge is placed. But with sufficient care, this can be dealt with, or pressure can be estimated by less direct means. In a particularly comprehensive study done for tidal flow around a headland in Puget Sound (e.g., Warner et al., 2013) it was found that the tidal topographic drag within the sampling region was about 30 times greater than the integrated bottom stress. This finding is consistent with the Foreman et al. (1995) Puget Sound tidal model that employed an improbably large bottom friction coefficient (about 5–10 times a normal drag coefficient value) to obtain reasonable results when lee waves were not resolved. Measurements such as Warner et al.'s are exceedingly uncommon (but see also Wijesekera et al., 2014), and so it is difficult to say how representative their findings are, but they most certainly tell us that form drag is a force to be reckoned with in tidal modeling and may well be an important contributor to global tidal dissipation.

## 7.8.    Conclusion

While tidal processes represent perhaps the oldest aspect of physical oceanography, the topic remains very lively. At the time of this writing, there are very active efforts involving internal wave generation, topographic drag, and dissipation. This work is attractive partly because of the great importance of related phenomena, such as global tidal dissipation (hence, even astronomical effects, such as changes in the Earth-Moon orbital parameters) or biological productivity, and partly because it provides a tool for understanding underlying processes. For example, Garrett (1972) was able to take advantage of precise tidal records to establish constraints on dissipation in the Bay of Fundy system, and Warner et al. (2013) were able to exploit the cyclic nature of tides to make better estimates of form stress at a ridge.

Tides are the great common denominator of the coastal ocean. They are present almost everywhere that the shelf adjoins a major ocean basin. In some cases, tides are a quiet background that primarily acts to enhance the bottom stress (see chapter 3), while in other cases, such as around the British Isles, tides and their consequences are by far the dominant physical processes. But they are always one of the fundamental descriptors of coastal physical processes.

# 8
# Freshwater Outflows

## 8.1. Introduction: Where Runoff from Land Encounters the Ocean

Outflows of relatively fresh water represent a major contact zone between our familiar terrestrial world and the ocean. The issue is not simply the passage of water but what can come with it: nutrients, sediments, and contaminants, for example. It is thus important to understand how fresh water moves about over the shelf and is eventually dispersed into the broader ocean. A particularly vivid example involves the outflow from the Mississippi River (Figure 8.1). The river often carries a high concentration of dissolved nutrients—enhanced by fertilizer usage, animal wastes, and products of fossil fuel combustion. When these nutrient-rich waters enter the ocean and move westward, high biological activity (phytoplankton growth) results. With time, the resulting organic material sinks, and as it decomposes, it consumes oxygen. During seasons of strong stratification, deep waters cannot be mixed to the surface, and subsequently near-bottom shelf waters become so oxygen depleted as to be inhospitable to animal life. This *hypoxia* (low-oxygen condition) is repeated every summer and is gradually growing in extent to the point that the 2019 "dead zone" was about the same size as the U.S. state of Massachusetts. The problem is not unique to the Mississippi outflow but occurs frequently in continental shelf waters globally (e.g., Rabalais et al., 2014).

The outflow problem is particularly interesting physically, because the density of the introduced waters is different from that of the ambient water, and so the former provide their own driving agency. While these density differences are often associated with lighter, low-salinity outflows, the density contrast might also be due to temperature (e.g., Masse and Murthy, 1990). Further, there are also "inverse estuaries," such as the Mediterranean Sea or Australia's Gulf of Carpentaria (e.g., Wolanski, 1986), where outflow water, made salty by evaporation, is denser than the ambient ocean waters. In either case, the horizontal density differences represent a pool of potential energy that can ultimately contribute to both current generation and mixing processes.

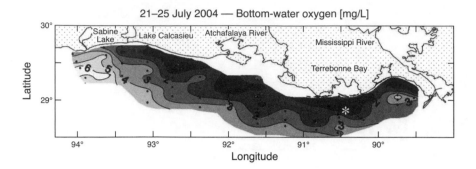

**FIGURE 8.1.** Hypoxia: observed dissolved oxygen depletion over the shelf. Near-bottom dissolved oxygen concentration over the Louisiana shelf during the summer of 2004. Darker shading corresponds to lower oxygen. The mouth of the Mississippi River is near 29°N, 89.3°W. Adapted from Rabalais et al. (2009).

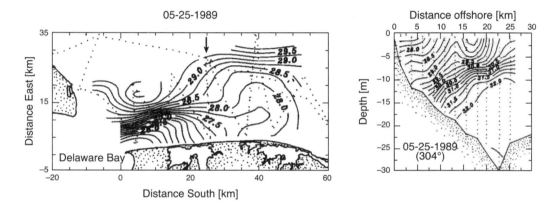

**FIGURE 8.2.** Frontal structures observed near an estuarine outflow. Surface salinity (left panel; north is to the left) and a cross section (right panel; taken on the line denoted by the arrow on the left panel) at the mouth of the Delaware estuary on May 25, 1989, a time when winds are minimal. Note (left panel) that the front at the outflow's edge appears to extend into the estuary, consistent with inflow from the shelf to the estuary occurring north of the outflow (rather than below the outflow). Adapted from Münchow and Garvine (1993).

Freshwater outflows can occur over a tremendous range of scales and settings. Further, the outflows are typically very time-dependent, so, all considered, there is no single typical case. Rather, outflows can be classified by the magnitude of the discharge from land, the importance of tides and wind driving, shelf geometry, and other factors to be discussed. A simple, purely linear approach to the problem suggests that whatever water comes out onto the shelf turns to the right (in the northern hemisphere) and spreads in the direction of coastal-trapped wave propagation (e.g., Figure 8.2). This follows because long coastal-trapped waves would presumably be instrumental in the current's initial formation. While this linear thinking is too simplistic in most cases, it does indeed capture the large-scale asymmetry correctly. This alongshore current, in turn, determines the scale over which a river's influence can be detected. There is no lack of interesting facets in the study of buoyancy currents on the shelf.

## 8.2.   Estuaries

An estuary is typically a nearly enclosed area where completely fresh river water first encounters intruding salty water and where mixing leads to a brackish outflow onto the continental shelf. Familiar examples abound, including the Hudson River (near New York City), the Chesapeake Bay, or a Norwegian fjord, but the variety is far broader than these few examples. A good deal of authoritative work has been done on estuaries (e.g., Mac-Cready and Geyer, 2009; Geyer and MacCready, 2014; Bruner de Miranda et al., 2017), and no attempt is made here to replicate these syntheses. Rather, a few essential properties, common to many cases, are described.

The classical image of an estuary has a completely fresh river flowing into a nearly enclosed area where saltier seawater penetrates. The fresher river water passes above the denser water that intrudes landward. The two water types mix, so the intrusion water becomes increasingly fresh landward, while the shallower, lighter outflow water becomes increasingly salty as it approaches the open shelf. The mixing of these two water masses is quantified by the well-known Knudsen relations, which simply describe the conservation of volume and of salinity in order to relate the incoming volume flux of fresh river water $Q_R$ to the outgoing flux of brackish water $Q_0$ (with salinity $S_0$) and the inward flux of salty shelf water $Q_S$ (with salinity $S_S$). Thus,

$$Q_0 = \frac{S_S}{\Delta S} Q_R, \tag{8.2.1a}$$

$$Q_S = \frac{S_0}{S_S} Q_0, \tag{8.2.1b}$$

where $\Delta S$ is the difference between the ambient shelf salinity entering the estuary and the outgoing brackish salinity. These relations do not close the problem, but they make an important point. Very often, the salinity $S_0$ of the water leaving the estuary (about 25 ‰ in Figure 8.2) is only a few units fresher than the ambient shelf water (about 31 ‰ in Figure 8.2), meaning that (from 8.2.1a) the outflow volume flux $Q_0$ has to be much greater than that of the incoming river $Q_R$ and that vigorous mixing occurs within the estuary. Further, the volume of shelf water flowing into the estuary is comparable to the volume of brackish water leaving (because $S_S$ does not differ too greatly from $S_0$). Two important sources of energy for this mixing are associated with (1) the interfacial shear between the outward-flowing shallower waters and the incoming waters, and (2) turbulence associated with tidal currents, which are often strong within estuaries because of the funneling geometry.

The preceding description is grossly simplified, of course, and ignores the variety of estuarine settings. For example, in wide (in the sense of mouth width compared with internal Rossby radius) estuaries such as Delaware Bay, the inflow can occur next to the outflow rather than under it (e.g., Münchow and Garvine, 1993; Figure 8.2). Also, model results demonstrate that lateral (cross-stream/vertical) circulation plays an important role within the estuary as well (Lerczak and Geyer, 2004). A range of conditions, then, including geometry, tidal amplitude, and river volume flux, combine to determine the outflow salinity $S_0$ and volume flux $Q_0$. Geyer and MacCready (2014) showed that the estuarine properties can be well characterized in terms of a single parameter that represents the ratio of a tidal time scale to a vertical mixing time scale.

Tides and buoyancy-driven flows are not the only important driving agencies within an estuary. Winds, both local and remote, can also affect conditions. For example, Wang and Elliott (1978) examined sea level records in Chesapeake Bay (a rather large estuary) and found that winds along the bay's axis drive fluctuations. Further, conditions on the shelf outside the estuary also have a profound effect on sea level inside the bay. Physically (Jackson et al., 2018), this shelf influence can be thought of as a wind-driven coastal-trapped wave approaching the bay from the north, changing sea level at the mouth, and thus requiring a response within the bay. Indeed, these ultimately wind-driven effects are so strong as to modulate the bay's outflow (e.g., Valle-Levinson et al., 2001) and thus make the subsequent alongshore buoyancy current pulse-like (e.g., Rennie et al., 1999).

At the actual outflow site, where the brackish water leaves the enclosed estuary, the scale of the flow, in the sense of a Rossby number $Ro = \hat{u} / (fL)$ (where $\hat{u}$ is a representative velocity, $f$ is the Coriolis parameter, and $L$ is a representative horizontal length scale), is critical. For large $Ro$, Earth's rotation is negligible, and flow tends to go directly offshore while fanning out and sometimes undergoing hydraulic behavior (Horner-Divine et al., 2015). When there is a rapidly flowing buoyant layer, shear across its lower interface can lead to substantial mixing in the near field before spreading decelerates the flow (e.g., MacDonald et al., 2007). Alternatively, slower (lower Rossby number) outflows are affected by Earth's rotation and are perhaps less susceptible to mixing in the near field.

## 8.3.    Generating an Alongshore Flow: The Bulge

Once the initial outflow from the estuary has spread out sufficiently, it slows down to subcritical (in the sense of a Richardson number: eqn. 2.3.7) shears, so interfacial mixing decreases. It is then a reasonable first approximation to neglect dissipation to understand what the outflow does next. Nof and Pichevin (2001) considered a very idealized case of upper-layer water flowing out from land into an ocean with $f > 0$, a very deep, quiescent lower/ambient layer and a vertical wall at the coast (Figure 8.3). They constructed a control volume (ABCD in the figure) large enough to include the entire adjustment region where the estuarine outflow turns and begins to move alongshore. By careful treatment of the alongshore ($y$) momentum balance of the upper layer water, they came to an interesting conclusion. The integrated (across AB) outward momentum flux $v^2$ in the alongshore buoyancy current is unbalanced unless there is a gradually growing bulge region within the control volume. Thus, not all the water flowing out from the land (through

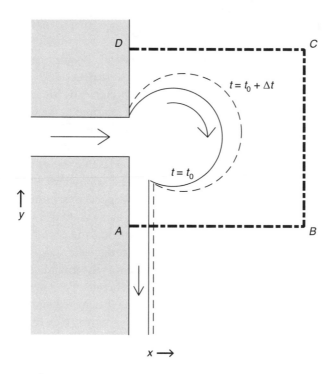

**FIGURE 8.3.** Schematic of bulge development as fresher water leaves an estuary in the northern hemisphere. The momentum flux out through boundary AB is balanced by the growing "balloon" of outflow water within the control volume. This growth is indicated by showing the edge of the bulge at two times.

section AD) proceeds alongshore in the buoyancy current (through section AB); some remains in the slowly expanding bulge.

More insight can be gained by assuming that the flow circulating around the bulge is in gradient wind balance (i.e., the radial pressure gradient is balanced by the sum of the Coriolis force acting on the swirl velocity $\check{v}$ and the centrifugal acceleration). Further, if the bulge is circular and moving in a rigid body rotation,

$$\check{v} = -\frac{\gamma f r}{2}, \tag{8.3.1}$$

(where $r$ is the radius within the bulge, and $\gamma$ is a constant $< 1$), then it can be shown (Nof and Pichevin, 2001) that the volume flux in the coastal current (i.e., outward through boundary AB) is given by

$$|q| = \frac{Q}{1+2\gamma}, \tag{8.3.2}$$

where $Q$ is the volume flux out of the estuary (through section AD). Although $\gamma$ may not be straightforward to find, this expression does make it clear that the stronger (more nonlinear) the flow is, the more outflow water goes into bulge growth, and the less goes into the alongshore coastal current. It is satisfying to note that in the linear limit, $\gamma \to 0$, all the estuarine outflow turns and flows alongshore.

For some time prior to Nof and Pichevin's (2001) study, people treating river outflows using numerical models or laboratory experiments were puzzled by bulge formation in their results. Because bulge structures had not yet been clearly identified in the ocean, people tended to regard the bulge as an artifact to be removed by, for example, including an ambient alongshore current (in the $-y$ direction in Figure 8.3). Indeed, once numerical models are made more realistic, the bulge is often less pronounced (e.g., Garvine, 2001). For example, if realistically sloping bottom topography or vanishing depth at the coast is included, the bulge weakens. Further, if the discharge from land is at an acute angle to the coastline (i.e., if the river comes from the northwest rather than the west in Figure 8.3), the bulge can again be weakened. Garvine also showed that the bulge spreads strongly in the northward direction if the discharge from land is simply given as an outflow rather than as a more realistic inflow/outflow condition.

An important step in understanding the bulge was the unambiguous detection of bulges in ocean observations. The problem is difficult because the ambient ocean is active and very time-dependent: tides and wind-driven currents, for example, advect the bulge and create very strong observational constraints on synopticity (i.e., you need to sample it quickly before it changes). In this sense, it is perhaps not surprising that an early, unambiguous observation of a growing bulge with a buoyancy current was made using satellite remote sensing (a truly instantaneous measurement) in Lake Ontario, where there are no tidal currents to complicate matters (Horner-Devine et al., 2008). Early in situ bulge observations in the ocean were made by Chant et al. (2008) in the Hudson River outflow, and by Horner-Devine et al. (2009) off the Columbia River mouth.

It has become clear that a growing bulge in the ocean offshore of an estuary's mouth is to be expected in many circumstances, although not in a form as stark as what Nof and Pichevin (2001) might have envisioned. Rather, effects such as frictional dissipation, bottom topography, and complex ambient currents make bulges in nature less pronounced and often transitory. This contrasts with many idealized numerical modeling studies, where a bulge often occurs when idealized conditions are applied (e.g., Fong and Geyer, 2002).

## 8.4.   The Buoyancy Current

Now that the alongshore buoyancy current's volume flux $q$ (flowing southward in Figure 8.3) has been established, its properties can be studied. To understand the scales and structure associated with this feature, it is useful to consider the *geostrophic adjustment* problem. In this highly idealized case, a motionless body of light water is adiabatically allowed to reach a geostrophic equilibrium that demonstrates the inherent scales. The initial buoyant layer (dashed line in Figure 8.4) adjoins a vertical coastal wall, and the lower layer is assumed to be deep and motionless. The upper layer's properties depend only on the cross-shelf distance $x$, and the adjusted alongshore flow is assumed to be geostrophically balanced (section 5.7), so that

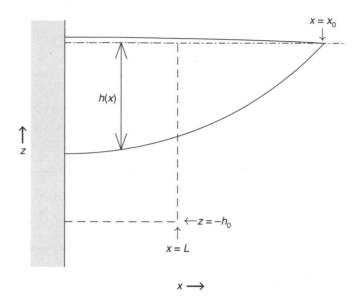

**FIGURE 8.4.** Schematic cross section of a buoyancy current. Dashed line: the interface depth of a motionless initial state. Solid curve: the geostrophically adjusted interface depth. The level dash-dot line, for reference, is $z = 0$. For this example, $x_0/\Lambda = 1.63$, and $L/\Lambda = 0.71$, where $\Lambda$ is defined in (8.4.4b).

$$fv = \frac{1}{\rho_1} p_x, \qquad (8.4.1a)$$

$$v_t + u v_x + fu = 0, \qquad (8.4.1b)$$

$$(hu)_x + h_t = 0. \qquad (8.4.1c)$$

Because the lower layer is at rest, the upper layer pressure gradient is

$$p_x = g' \rho_1 h_x, \qquad (8.4.1.d)$$

where $(u, v)$ are the upper layer cross-shelf and alongshore current components, respectively; $\rho_1$ and $\rho_2$ are the upper- and lower-layer densities; $h$ is the thickness of the upper layer; and $g'$ is the reduced gravity $g(\rho_2 - \rho_1)/\rho_2$. Equations (8.4.1b, c) can be combined to show that *potential vorticity* following a water parcel is conserved:

$$\frac{d}{dt} \frac{f + v_x}{h} = 0, \qquad (8.4.2a)$$

where $\dfrac{d}{dt}$ is the total derivative

$$\frac{d}{dt} = \frac{\partial}{\partial t} + u \frac{\partial}{\partial x}. \qquad (8.4.2b)$$

Now, consider an initially motionless body of water, with depth $h_0$ and width $L$ (dashed line in Figure 8.4). The problem then is to find the eventual distribution of layer depth and velocity once a geostrophic equilibrium is reached, assuming that vorticity is conserved. Thus, initial potential vorticity of a water parcel equals the final value, so that

$$\frac{f}{h_0} = \frac{f + v_x}{h},$$

(8.4.3)

where $h(x)$ is the adjusted depth of the interface between the two layers, and $v(x)$ is the adjusted geostrophic alongshore velocity. The solution to (8.4.3), given (8.4.1a) and (8.4.1d), can then be written as

$$h = h_o + C_1 \cosh(x/\Lambda) + C_2 \sinh(x/\Lambda),$$

(8.4.4a)

where $\Lambda$ is the internal Rossby radius of deformation,

$$\Lambda = \sqrt{g'h_0}/|f|.$$

(8.4.4b)

The coastal boundary condition follows from the alongshore momentum equation (8.4.1b). Because there is no flow through the wall ($u = 0$ at $x = 0$), $v$ at the coast never changes from its initial rest state. Thus, $v = 0$ at $x = 0$, so that (from 8.4.1a) $h_x = 0$ at $x = 0$, and thus $C_2 = 0$. Also, define $x_0$ as the location where the adjusted depth vanishes (see Figure 8.4) and require that upper layer volume is conserved:

$$h_0 L = \int_0^{x_0} h \, dx,$$

(8.4.5)

so

$$C_1 = \frac{-h_0}{\cosh(x_0/\Lambda)},$$

(8.4.6a)

and $x_0$ is found by solving

$$\frac{L - x_0}{\Lambda} = -\tanh(x_0/\Lambda).$$

(8.4.6b)

The adjusted interface depth is displayed as a solid line in Figure 8.4. The final depth at the coast, $h_W = h(0)$, is less than the initial depth $h_0$, and $x_0 > L$ because of the way the upper layer spreads out as a lens. Finally, the alongshore transport of the adjusted current is

$$q = \int_0^{x_0} v h \, dx = \frac{g'}{f} \int_0^{x_0} h h_x \, dx = -\frac{g' h_W^2}{2f},$$

(8.4.7a)

so the depth at $x = 0$ is

$$h_W = \sqrt{\frac{2|f q|}{g'}}.$$

(8.4.7b)

These results are admittedly very idealized, but they do introduce the point that the internal Rossby radius (8.4.4b) is the natural cross-shelf scale, as well as the simple relations (8.4.7), which will prove remarkably durable so long as the alongshore flow is geostrophically balanced.

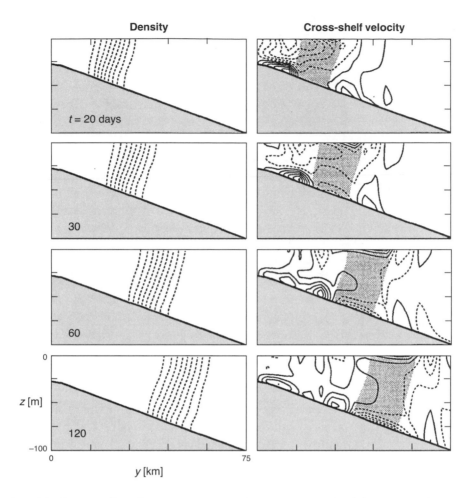

**FIGURE 8.5.** Adjustment of frontal location so as to minimize bottom stress. Evolution of salinity in a model buoyancy current (left panels, where fresher water is at the coast), and offshore velocity (right panels: offshore flow has solid contours, onshore is dashed, and the frontal area is shaded). Different rows represent different times. Adapted from Chapman and Lentz (1994).

The preceding model has at least two important shortcomings relative to actual ocean conditions: (1) it has a vertical wall rather than a sloping bottom, and (2) it neglects frictional processes. Chapman and Lentz (1994) addressed both of these shortcomings and arrived at strikingly clear results. They used a simple numerical model with a uniform bottom slope, introduced a freshwater input, and let the model adjust downstream. Initially, southward (in the sense of Figure 8.3) flow is distributed in a region bounded by a salinity/density front (Figure 8.5, first panel). Inshore of the front, the southward flow induces an offshore bottom Ekman transport that helps move the front offshore. This adjustment continues until the alongshore velocity inshore of the front becomes very weak and the bottom Ekman transport becomes negligible. In this adjusted state, to a reasonable approximation, $v = 0$ inshore of where the front intersects the bottom, and

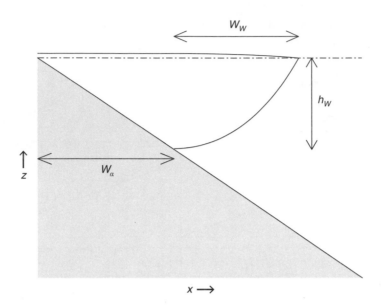

**FIGURE 8.6.** Schematic cross section showing the density interface for a frictionally adjusted buoyancy current over a sloping bottom. $W_\alpha$ is the distance from shore where the water depth is $h_W$ (8.4.7), and $W_W = x_0$ from the vertical wall case.

the entire alongshore transport is concentrated above the sloping front. Because the near-bottom velocity is zero at the front, frictional effects become negligible. It thus becomes reasonable to approximate this state as in Figure 8.6: an interface intersecting the bottom, and all alongshore flow concentrated above the sloping interface. The alongshore transport is given by (8.4.7), so that, for a given transport $q$, the interface intersects the bottom where the water depth equals $h_W$. Thus, the simple, vertical wall result—where the interface intersects solid ground—carries over to a far more realistic setting. Although bottom Ekman transport acts to eliminate bottom stress from the problem, this process differs somewhat from the buoyancy arrest addressed in section 3.7. The distinction is that here alongshore flow in the entire water column is affected rather than only within the bottom boundary layer (Chapman, 2002).

The idealizations embedded in Figure 8.6 make it possible to estimate the speed with which the buoyancy current develops (Lentz and Helfrich, 2002). Say the flow out of an estuary begins suddenly. Then, after initiation of a bulge, the alongshore current begins to flow, but it does not do so instantaneously. A "nose" moves forward into the motionless ambient waters and leaves behind it a structure like that of the figure even though flow near the nose can be disorderly and turbulent. Now, consider the nose more closely. Over a time $\Delta t$, the nose moves forward (southward in Figure 8.3 when $f > 0$) by a distance $\Delta y$. Propelling this motion is an alongshore-transported volume of $|q|\Delta t$. This lighter water, transported in the more offshore part of the freshwater domain (i.e., where there is a sloping interface), turns onshore and fills out a space that has the $(x, z)$

cross-sectional area $\Xi$ of lighter water in Figure 8.6 (approximating the interface as a straight line, rather than a curving one):

$$\Xi \approx \frac{1}{2} W_\alpha h_W + \frac{1}{2} W_W h_W. \tag{8.4.8a}$$

Given the constant bottom slope $\alpha$, $W_\alpha = h_W/\alpha$, and the authors estimated $W_W$ as a Rossby radius

$$\frac{\sqrt{g' h_W}}{f} \equiv \frac{c_W}{f} \tag{8.4.8b}$$

(note that this definition differs from equation 8.4.4b because it uses the adjusted depth $h_W$). Thus,

$$\Xi \approx \frac{1}{2} \left( \frac{h_W^2}{\alpha} + \frac{c_W h_W}{f} \right). \tag{8.4.8c}$$

Because the volume of water imported in time $\Delta t$ has to match the expansion $\Xi \Delta y$, the nose moves with a speed

$$\frac{|\Delta y|}{\Delta t} \equiv c_N = \frac{2|q|}{\dfrac{h_W^2}{\alpha} + \dfrac{c_W h_W}{f}}. \tag{8.4.9}$$

Using (8.4.7) and (8.4.8b), this simplifies to

$$c_N = \frac{c_W}{1 + \dfrac{c_W}{c_\alpha}} = \frac{c_W c_\alpha}{c_W + c_\alpha}, \tag{8.4.10a}$$

where

$$c_\alpha = \frac{g' \alpha}{f}. \tag{8.4.10b}$$

The latter equality in (8.4.10a) (analogous to the electrical circuit formula for the net resistance of a pair of parallel resistors) makes it clear that the speed $c_N$ can be no larger than the smaller of $c_W$ or $c_\alpha$. In the limit of a very steep bottom slope (approaching a vertical coastal wall), $c_\alpha \to \infty$, the nose moves with the speed of a gravity wave bore, $c_N \cong c_W$, but for a very gently sloping bottom (where the interface intersects the bottom far from shore), $c_N \cong c_\alpha$, and the nose moves very slowly. Lentz and Helfrich demonstrated the applicability of (8.4.10) using both laboratory experiments and numerical model results. Careful comparison against field observations (e.g., Washburn et al., 2011; McSweeney et al., 2021) shows that there is typically a good deal of scatter around the idealized result (8.4.10), evidently owing to effects such as ambient currents, mixing, and topographic variability.

To this point it has been assumed that the ambient ocean is at rest. Clearly, this cannot be the case, because there must be a salty inflow to the estuary for the outflow water to have some nonzero salinity. Where this inflow water comes from is important in terms of defining water properties, such as nutrient concentrations, in the estuary. The inflow water was treated carefully by Brasseale and MacCready (2021) by means of a sequence of idealized numerical models of an estuarine outflow into a homogeneous ambient ocean. One common thread in their calculations is that much of the inflow toward the estuary arrives from the south (in the sense of Figure 8.3, that is, the direction toward which long coastal-trapped waves propagate)—a result consistent with the notion that coastal-trapped waves play an important role in establishing the inflow pathway. Further, some of this water passes beneath the outflow's frontal zone, that is, in the lower layer between $x = W_\alpha$ and $x = W_\alpha + W_W$ (Figure 8.6). This northward lower-layer flow increases the vertical shear across the interface and so requires that the front's bottom position move farther off-shore and that the frontal zone widen. Additional transport toward the estuary occurs offshore of the frontal zone in a manner qualitatively resembling an "arrested topographic wave" (section 5.11). Finally, especially when the bottom slope is weak, some water enters the estuary from the near field on the northern side (in Figure 8.3, that is, from the direction toward which short barotropic shelf waves propagate) of the estuary.

## 8.5.    Wind Forcing

Both wind forcing and buoyancy drive alongshore flows, so it is useful to know the relative strength of the two mechanisms and how the two forcings interact. One might estimate the wind-driven current magnitude by assuming that alongshore surface stress equals bottom stress, so that a representative current $\widehat{v_W} = \tau_0^y / (\rho_0 \sigma_F)$, where $\tau_0^y$ is the alongshore wind stress, and $\sigma_F$ is a frictional bottom resistance parameter (chapter 3). A representative alongshore flow within a buoyancy current (from 8.4.1 and assuming the lower layer is at rest) might be about $\widehat{v_B} = g'H / (f L)$, where $H$ is the water depth, and the horizontal length scale is about an internal Rossby radius, so that $\widehat{v_B} = \sqrt{g'H}$. Thus, a typical wind-driven current versus a typical buoyancy current would be about

$$\frac{\widehat{v_W}}{\widehat{v_B}} = \frac{\tau_0^y}{\rho_0 \sigma_F \sqrt{g'H}}. \tag{8.5.1}$$

While this ratio[1] provides a crude measure, the actual interactions of the two forcings are interesting in their own right.

The response of a buoyancy current to upwelling-favorable alongshore winds was summarized by Lentz (2004), who used physical reasoning and who evaluated his results with numerical models and observations. In this case, the wind stress opposes the buoyancy current, and the surface Ekman transport strongly distorts the buoyancy current's structure.

Consider an initial buoyancy current flowing along a vertical wall, and, for simplicity, assume that the interface slopes linearly ($t = 0$ solid line in Figure 8.7) as opposed to

---

[1] The reader might also consult Whitney and Garvine (2005) for an alternative scaling based on somewhat different assumptions.

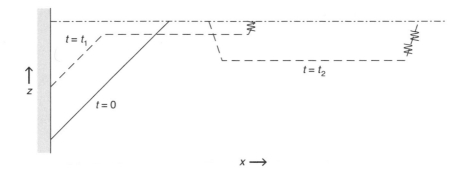

**FIGURE 8.7.** Schematic cross section showing stages of the spreading and dilution of a buoyancy current under the influence of an upwelling-favorable alongshore wind stress. The solid line is the initial state. The zigzag lines in the subsequent states indicate sites of enhanced vertical mixing.

the more realistic curved shape (Figure 8.4). The ambient ocean is motionless and homogeneous, having density $\rho_2$. When a steady, upwelling-favorable wind is applied, the Ekman transport $U_E = \tau_0^y / (\rho_0 f)$ advects surface waters offshore within a turbulent boundary layer and so stretches out the cross-sectional area, reducing the size of the core current adjoining the coast ($t = t_1$ in Figure 8.7). The thickness of the extruded layer is taken to be governed by a bulk Richardson number criterion (see section 3.5), and entrainment can occur (if needed) near the offshore-penetrating "nose" of the current. This entrainment increases the density of the plume water while also increasing its total volume. In the limit of a small density difference between the evolved plume and ambient water, the offshore tongue can actually be thicker than the initial buoyancy current. If the wind forcing continues long enough, the entire diluted buoyancy current separates from the coast ($t = t_2$ in Figure 8.7) and moves offshore with the Ekman transport. The time scale for this reconfiguration process is given by $2\Xi / (\sqrt{R_B} U_E)$, where $R_B$ is a critical bulk Richardson number (a constant), and $\Xi$ is the initial cross-sectional area of the current. Regardless of the details, upwelling-favorable winds act to dilute the plume, to expand its cross-sectional area, and to spread it offshore. The net effect is that the plume tends to lose its identity as a distinct freshwater body.

When the winds are downwelling favorable, alongshore currents are enhanced by the added forcing, and the surface Ekman transport is shoreward. Moffat and Lentz (2014) treated this problem in an idealized manner and evaluated their results using primitive equation numerical model results. One limit occurs when the initial current is bottom advected (in the notation of Yankovsky and Chapman, 1997); that is, the frontal region is not wide, so that $W_\alpha \gg W_W$ in Figure 8.6. When the wind is first applied, there is some mixed layer deepening, and the onshore Ekman transport makes the frontal boundary more nearly vertical (Figure 8.8, left panel). A second limit occurs when most of the initial plume water is not in contact with the bottom—the front is surface advected; that is, $W_\alpha \ll W_W$. In this case, initial wind-driven mixed-layer deepening dilutes the plume and brings it closer to the bottom, if not all the way there. Further, the onshore surface Ekman transport again acts to make the front more nearly vertical. The net effect is to convert

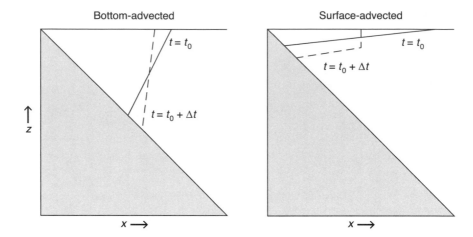

**FIGURE 8.8.** Schematic cross sections of a buoyancy current responding to a downwelling-favorable alongshore wind stress. Left: a bottom-advected front, where the primary effect of the wind stress is to make the front more nearly vertical. Right: a surface-advected front, where the wind stress leads to both layer deepening and to moving the surface expression shoreward.

the surface-advected front into a more nearly bottom-advected form (Figure 8.8, right panel). In either limit, once the adjustment toward a vertical front takes place (on a time scale similar to that of the upwelling case), there is a continued slow dilution of the buoyancy current by ambient waters, but the primary downwelling circulation tends to occur offshore of the front, leaving the plume waters relatively isolated. The situation is comparable to what happens with wind-driven downwelling in the absence of a buoyancy current (section 4.6; Figure 4.5; Austin and Lentz, 2002), but now the downwelling front is particularly strong because of the initial buoyancy current. Further, there can actually be an upwelling cell inshore of the front. In summary, downwelling-favorable winds accelerate the alongshore current but may not be as effective at diluting the buoyant waters as are upwelling-favorable winds.

## 8.6.   Turning It Upside Down: Negative Buoyancy

There are many examples of waters in the coastal ocean being denser than the surroundings. These might occur, for example, because of wintertime cooling over the shelf or outflow from a region, such as a neighboring sea, where evaporation creates saltier waters. In any case, the dense water often spills from the shelf and downward across the slope, tending toward an equilibrium, density-neutral, depth (e.g., Ivanov et al., 2004, for a collection of observed examples). A major difference between dense and lighter outflows is that the former are found in contact with the bottom and so are often exposed to frictional effects, while a buoyant current is usually idealized as unaffected by bottom stress at lowest order. The bottom drag, in turn, makes the behavior of negatively buoyant currents radically different from that in the light water case, where (away from an adjust-ment region; section 8.3) currents tend to hug the coast.

One useful idealization involves a steady point-source inflow onto the shelf or upper slope of the sort associated with the deep, dense flow from the Mediterranean Sea into the Atlantic Ocean. After perhaps an initial adjustment where nonlinear and/or dissipa-tive terms dominate, the dense, bottom-hugging current passes from the shelf onto the continental slope. On the slope the current flows in the direction of coastal-trapped wave propagation (just as would a buoyant outflow near the surface). The alongshore current may be in approximately geostrophic balance, where the cross-isobath pressure gradient is associated with the density difference between the outflow water and the ambient ocean. The cross-isobath current is strongly affected by the along-isobath bottom stress, giving rise to a downslope Ekman transport, which, in turn, forces the current toward greater depths. In addition, entrainment of ambient water across the current's boundaries decreases the density contrast between the current and its surroundings, thus decreasing the cross-isobath pressure gradient and weakening the along-isobath flow. In this idealiza-tion, the outflow water will eventually come to rest at a depth consistent with its adjusted (accounting for the entrainment) density, which is less than that of the initial, injected current. This idealization appears to work well for Gibraltar outflow waters (Baringer and Price 1997a, b), where the outflow core passes downward and poleward, and the flow appears to be, over about 200–300 km, relatively unperturbed by eddies (Bower et al., 1997).

This idealization has its limits, however. The Gibraltar outflow eventually leads to eddy formation (e.g., Bower et al., 1997), an inherently time-dependent and nonlinear process. After the Mediterranean outflow proceeds poleward and downward to about 1000 m depth, it breaks off about 15–20 times per year to form long-lived, relatively salty lenses ("meddies") that drift generally offshore, westward across the Atlantic. Typically, these features are about 200–1000 m thick and have a radius of about 10–50 km. A representa-tive swirl velocity is around 0.3 m/s, and it rarely extends all the way to the ocean's surface. Indeed, Jiang and Garwood (1996) carried out numerical model calculations indi-cating that point-source dense currents can often break up into discrete structures. One classic example of eddy variability, perhaps bearing more resemblance to Jiang and Gar-wood's result than to meddy formation, is found in the Denmark Strait overflow water along the continental slope off southeastern Greenland (e.g., von Appen et al, 2014). Here, along the 900 m isobath, energetic eddies are detected on a 2–4 day repeat cycle while propagating equatorward at about 0.7 m/s. These bottom-intensified features typi-cally involve 300 m-thick lenses of dense water that are about 8 km in radius and have swirl velocities of about 0.2 m/s near the bottom while remaining detectable even at the surface.

In many cases, a point-source approximation is not reasonable, as when there is dis-tributed dense water formation driven by surface cooling over the shelf, e.g., Gawarkiewicz (2000); Yankovsky and Legg (2019). A spatially uniform cooling causes surface-to-bottom mixing and creates the densest water in shallow areas but less dense in deeper locations. A geostrophic adjustment can occur, and the cross-shelf density gradient is susceptible to baroclinic instabilities (see section 9.8), As a result, intense eddies with horizontal scales $O(10 \text{ km})$ form in models and drift offshore. Once these reach the continental slope, they translate directly downward toward an equilibrium depth with little tendency to translate along isobath. Indeed, given the modeled size and nonlinearity of the individual eddies, one would not expect the sort of linear dynamics associated with setting up a steady along-isobath flow (e.g., section 5.11).

The Gibraltar and Denmark Strait outflows contrast in striking ways, such as eddy repeat times and the path of eddies once formed. A range of factors could be involved. including the speed of the underlying dense current, bottom slope, ambient stratification, or the orientation of the coastline. In contrast with the case with a prescribed inflow of dense water, distributed densification on the shelf can give rise to an eddy field that includes discrete dense structures sliding off the shelf and relatively directly offshore. Further, it seems possible that other configurations out of the Ivanov et al. (2004) catalog might be either stable or unstable in still other ways. In many cases, dense currents over the shelf and slope can be hard to observe owing to remoteness, intermittency, or foul weather. Yet, these issues are important because the downward motion of dense water is a critical part of the global ocean circulation: it is the pathway by which dense, high-latitude waters sink to an equilibrium depth, where they spread across the globe (e.g., Talley et al., 2011). Much remains to be learned about this class of phenomena.

## 8.7.    Conclusion

Buoyant outflows and their consequent shelf currents occur on a staggering range of scales. At the small end, outflows are affected by surf-zone physics (e.g., Rodriguez et al., 2018), and rotational effects are negligible. At the largest scales (such as the Middle Atlantic Bight shelf break jet (e.g., Linder and Gawarkiewicz, 1998) or the Norwegian Coastal Current (e.g., Johannessen et al., 1989) buoyancy currents are a major part of regional shelf circulation, and the physics is dominated by the effects of rotation. In all these cases, the primary question often reduces to the dispersion of the buoyant waters alongshore and across the shelf. Thus, questions of buoyancy current extension and extinction are critical even though mechanisms may be different at different scales. For example, upwelling-favorable wind forcing dissipates a buoyancy current on time scales of order $\Xi/U_E$ (cross-sectional area divided by Ekman transport), so it is most effective for relatively small-scale currents. This raises the question of whether other mechanisms, such as baroclinic instability (section 9.6), might not be more effective for larger-scale currents.

Buoyancy currents represent a very lively area of research. The advent of better observational tools allows more comprehensive and synoptic ocean measurements than ever in the past. At the same time, more powerful computers, insightfully used, are leading to more sophisticated dynamical understanding of the processes involved. Exciting advances, such as determining the role of realistically variable topography (e.g., Suanda et al., 2016) are still waiting to be made.

# 9

# Instabilities

## 9.1.  Introduction: Growth of Disorder over the Shelf

Hydrodynamic instability is a process by which an orderly flow configuration begins to break down and form (usually) smaller-scale features and eddies (e.g., Figure 9.1). The process extracts energy from the initial configuration and, very often, leads to the development of a disorderly eddy field. Instability plays such a strong role in our everyday lives that people may not even realize how commonplace it is: examples range from changes in weather to flow in our kitchen sinks.

Instability in the coastal ocean is important for a number of reasons. One is simply to be able to understand observed features in the ocean, although well-resolved eddy observations over the shelf are rare because of short time and space scales. However, as will be seen (section 9.4), the results of instabilities are evidently detectable indirectly by statistical methods. But perhaps more important, in many contexts, eddies can collectively transport mass, momentum, or dissolved materials, and this possibility needs to be evaluated. For example, in some given context, is cross-shelf eddy heat transfer comparable to that associated with relatively steady, larger-scale flows?

This chapter treats various forms of instabilities in the coastal ocean and, where possible, the consequent eddy field evolution (e.g., Figure 9.1). Over the shelf proper, the resulting eddies typically have horizontal scales of 2–10 km and are thus large enough that Earth's rotation is relevant. Within the surf zone, eddy scales are smaller, and Earth's rotation is less relevant. Emphasis here is placed on eddies in which motions are predominantly horizontal; genuinely three-dimensional turbulence is not treated here. Isotropic turbulence is, of course, extremely important in the coastal ocean, but it is treated very well elsewhere (e.g., Thorpe, 2007; Gregg et al., 2018; Moum, 2021; Gregg, 2021).

## 9.2.  A Framework

There are a number of excellent books and papers that treat ocean instabilities (e.g., Stone, 1966; Pedlosky, 1979; Gill, 1982; Haine and Marshall, 1998) in general contexts. This

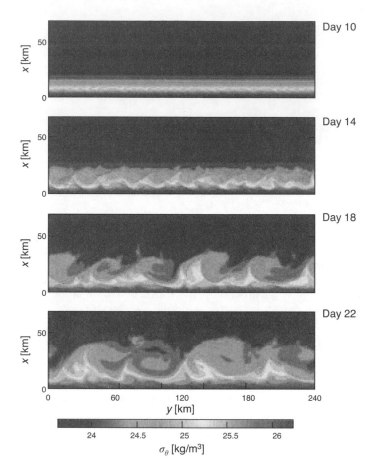

**FIGURE 9.1.** The evolution of a growing instability. Plots of surface density at different times from a numerical experiment showing how instabilities of an initial alongshore flow near $x = 0$ grow and become more chaotic with time. Lighter shades usually represent denser water. Time increases from the upper plot to the lowest, and the horizontal axis is alongshore. Adapted from Durski and Allen (2005).

chapter instead focuses on types of instabilities that should be found in the coastal ocean, as well as on their distinguishing context and properties. As background, much important work on instability theory takes advantage of the quasigeostrophic approximation (e.g., Blumsack and Gierasch, 1972). This assumption (which assumes moderate Rossby number, modest vertical isopycnal excursions, and small fractional changes in water depth), however, is rarely quantitatively applicable in the coastal context because of large depth changes and because very often isopycnal displacements occur over vertical scales comparable to the water depth. Most certainly, insights gained using quasigeostrophy are helpful, but they should usually be treated as qualitative findings in a coastal setting.

Instability typically involves extracting energy from an initial flow or density configuration so that smaller scale features can grow. In the present context, there are two general types of energy sources. One relates to the momentum of the initial flow field. For example, say the initial flow is a jet:

$$v_0 = C[1 - \cos(2\pi x/W)], \tag{9.2.1}$$

for $0 < x < W$, $x$ being the transverse coordinate. This configuration has an $x$-averaged kinetic energy per unit mass of $KE = (3/4)C^2$. The potential energy associated with the free-surface displacement is negligible if the channel is narrow relative to the Rossby radius of deformation, that is, if $f^2 W^2/(gH) \ll 1$, where $H$ is the water depth. Say, however, that one could rearrange the flow (putting aside the question of mechanism) while conserving total momentum, to a constant $v_0 = C$. Then, the averaged kinetic energy per unit mass would be $KE = (1/2)C^2$. Thus, it is conceivable to free up an energy per unit mass of $(1/4)C^2$ simply by redistributing the flow. A similar argument could be made for flow that varies in the vertical direction. Thus, a sheared flow can potentially shed some of its kinetic energy and still conserve momentum. But how could this actually happen, under what (if any) circumstances, and where does the liberated energy go?

Similarly, consider the gravitational potential energy (per unit mass) associated with a tilting interface [at $z = -H/2 + G(x - W/2)$] between fluids of density $\rho_1$ and $\rho_2$ in a channel of total depth $H$ and width $W$ (and $\rho_0$ is the mean density). In this case, the tilted interface case has

$$\Delta PE = g(\rho_2 - \rho_1)G^2 W^2 / (24 H \rho_0) \tag{9.2.2a}$$

more integrated potential energy per unit mass than when the interface is flat ($G = 0$). Assuming that the geostrophically balanced upper- and lower-layer velocities are equal and opposite, the initial kinetic energy per unit mass is

$$KE_0 = \frac{1}{8} \frac{g'^2 G^2}{f^2}, \tag{9.2.2b}$$

(where $g' = g(\rho_2 - \rho_1)/\rho_2$), and the final kinetic energy, with a flattened interface, is zero. The kinetic energy is negligible compared with $\Delta PE$ if the channel is wide compared with the *internal* Rossby radius, that is, if $f^2 W^2/(g'H) \gg 1$. It may not be possible to release even a portion of the mean state energy, but release may be possible in some conditions. For example, if the channel were not rotating, the interface would be free to slump, leading to a back-and-forth sloshing across the channel. However, if the channel were rotating, the tilt could be geostrophically balanced, and so slumping and sloshing do not appear to be an option. But at the same time it seems reasonable that perhaps there are extreme interfacial slopes that, even in a rotating system, would be somehow unsustainable as a steady state. Indeed, this possibility, in which gravitational potential energy associated with the density configuration is the main source of perturbation energy, represents the well-known *baroclinic instability*.

So, there are situations in which energy is nominally available to allow flow transitions, namely, instabilities. The next questions involve, first, whether the transition is possible in a given context and, second, what form the new motions would take. As in the case of the sloping interface, energy availability does not necessarily mean that instabilities can grow. These questions motivate the study of *linear stability*, which addresses whether and under what conditions small disturbances can grow, thus exploiting the initial energy pool. The methodology of this approach is to start with a given flow, linearize the equations of motion about that configuration, solve for possible disturbances, and ask under what conditions they will grow in time or space. This is a powerful and illuminating methodology, but it is valid only for the earliest phases of disturbance growth, when the initial flow has not yet changed substantially. It then becomes necessary to ask how the growing disturbances evolve as the initial flow changes. Disturbances are not expected to grow indefinitely, because the initial state evolves with time to a state having less mean energy. Further, as disturbances become eddies, and the eddies interact, the eddy properties change. For example, it is fairly common for a field of eddies on a rotating planet to evolve so that individual eddies merge and expand with time (e.g., Rhines, 1977, in the quasigeostrophic limit). This inherently nonlinear evolution is often tackled using numerical models informed by physical reasoning, which helps in creating scalings for properties of the eddy field.

One useful way to study both instability and eddy evolution is to quantify the energy forms and the transfer of energy from one pool to another. Here, and for the remainder of this chapter, hydrostatic conditions (horizontal scales much larger than vertical) are assumed. For the purpose of illustration, consider an alongshore-cyclic (in $y$) channel, with boundaries at $x = 0$, $W$, and depth $h(x)$. Averaging along the channel yields, for example,

$$q = \{q\} + q^\dagger, \tag{9.2.3}$$

where $\{q\}(x, z, t)$ is the along-channel mean of $q$ and $q^\dagger(x, y, z, t)$ is the deviation from that mean. There is then a *mean kinetic energy* per unit mass

$$MKE = \frac{1}{2\Xi} \int_0^W \int_{-h}^0 [\{u\}^2 + \{v\}^2] dz\ dx, \tag{9.2.4}$$

a *potential energy* per unit mass

$$PE = \frac{1}{\rho_0 \Xi} \int_0^W \int_{-h}^0 \{g\rho z - g\rho_0 z_0\} dz\ dx, \tag{9.2.5}$$

and an *eddy kinetic energy* per unit mass

$$EKE = \frac{1}{2\Xi} \int_0^W \int_{-h}^0 \{u^{\dagger 2} + v^{\dagger 2}\} dz\ dx, \tag{9.2.6}$$

where $\Xi$ is the cross-sectional area of the channel, $\rho_0$ is a constant reference density, and $z_0$ is an arbitrary reference level. The *PE* associated with the free-surface tilt has been ignored in (9.2.5), with the understanding that the channel is narrow compared with the barotropic Rossby radius (i.e., that the rigid-lid approximation is valid). A complete energy analysis would, of course, also include terms representing external forcing, energy loss (both mean and eddy) due to frictional effects, and energy transports (advective, eddy, and radiative) through any open boundaries.

Of greatest interest are the conversions among these energy pools. One conversion is between potential energy and kinetic energy:

$$C_{PE \to KE} = -\frac{g}{\rho_0 \Xi} \int_0^W \int_{-h}^0 [\{w\}\{\rho\} + \{w^\dagger \rho^\dagger\}] dz \, dx, \qquad (9.2.7)$$

which consists of two parts. First, the $\{w\}\{\rho\}$ term represents a loss of potential energy associated with an alongshore-uniform slumping such as would be expected in the non-rotating channel with an initially tilted interface. The second term, $\{w^\dagger \rho^\dagger\}$ represents the change of potential energy associated with motions that vary along the channel. Whether these motions are wavelike or eddies, the term is referred to as an "eddy conversion". Baroclinic instability draws its energy through this transformation. Further, it is often useful to define the *available potential energy APE* as the actual potential energy (9.2.5) with tilting isopycnals minus the potential energy if all particles retain their density but are re-arranged so that isopycnals are flat. This *APE* represents an upper bound to the amount of eddy energy that could be created from a given density configuration.

A second important conversion is that from *MKE* to *EKE*. Specifically,

$$C_{MKE \to EKE} = -\frac{1}{\Xi} \int_0^W \int_{-h}^0 [\{v_x\}\{u^\dagger v^\dagger\} + \{u_x\}\{u^\dagger u^\dagger\} + \{v_z\}\{w^\dagger v^\dagger\} + \{u_z\}\{w^\dagger u^\dagger\}] dz \, dx. \qquad (9.2.8)$$

The terms in this conversion all have forms like $\{v_x\}\{u^\dagger v^\dagger\}$, which represent spatial variations of the mean velocity $\{v_x\}$, and an eddy flux $\{u^\dagger v^\dagger\}$. The first two terms in (9.2.8) represent the process of *barotropic instability*, which extracts energy associated with the horizontal current shear (as in eqn. 9.2.1) and thus smooths out the initial horizontal variations in along-channel mean flow. (Note that the term "barotropic" here does not require that the flow be barotropic for the instability to grow but only that the horizontal shear be the energy source). The second pair of terms (that involve $w^\dagger$) represent energy release due to the weakening of vertical shears: an exchange that is negligible in quasi-geostrophic theory but that can be important near features such as fronts.

The energy conversions quantified by (9.2.7–9.2.8) illustrate another point as well. Specifically, these formulations can have blind spots. In this case, a potentially important instability will be missed: *symmetric instability* (e.g., Allen and Newberger, 1998), which can manifest as alongshore-uniform rolls that form when lateral shears (relative to tilting density surfaces) exceed a critical value (e.g., Haine and Marshall, 1998). Because of the $y$ uniformity, $u^\dagger$ and $w^\dagger$ are zero, and all motions tied to the instability are hidden within $\{u\}$ and $\{w\}$. Thus, even when the instability is growing, *EKE*, as defined by (9.2.6) remains zero. This shortcoming can be remedied by defining the mean as being in $x$ or $t$ rather than in $y$, but these alternative definitions can lead to other difficulties.

There are many situations in the coastal ocean that can, at least in principle, give rise to instabilities. Rather than trying to do equal justice to all these possibilities, a few illustrative cases are singled out, and others are treated in less detail than they might deserve. Some of these examples have been investigated only with models, and detailed observational verification is often absent. This observational scarcity can be attributed to several causes. One is that on many continental shelves, other forms of current variability, such as tides or wind-driven alongshore flows, are the dominant features and thus complicate the study of sometimes-weaker eddy fields. Another important difficultly arises from the scales often associated with eddies in the coastal ocean. For example, the eddies driven ultimately through wind forcing (Figures 9.1 and 9.2; sections 9.3 and 9.4) have modeled length scales of about $O(10 \text{ km})$ or less and Eulerian time scales of hours. Mobile features with scales like these are challenging to observe, and, in fact, the well-observed examples that do exist call for very concentrated efforts.

## 9.3.    Coastal Upwelling Front

In coastal upwelling systems, alongshore winds drive offshore surface Ekman transport, which is compensated for by an onshore and upward flux of deeper, colder, nutrient-rich water (see chapter 4). In many cases, the surfacing of isopycnals creates a front (e.g., Figure 4.2) which is typically associated, through geostrophy, with an alongshore jet of a few tenths of a meter per second. There has been evidence since the 1970s, in the form of sea surface temperature maps and satellite imagery, that these fronts are unstable, with a wavelength of perhaps 40 km (e.g., Holladay and O'Brien, 1975). The stability problem was tackled thoroughly by Barth (1994), who started with an idealized representation of the upwelling front off Oregon (similar to the temperature section in Figure 4.2), and demonstrated that it is linearly baroclinically unstable, with two unstable modes. He then used a primitive equation numerical model to demonstrate the growth of the unstable disturbances into eddies. The eddies themselves undergo an inverse scale cascade (e.g., see Rhines, 1977, for the quasigeostrophic limit of this process), so the typical eddy size increases with time (as in Figure 9.1, which is meant to be representative of Oregon summertime conditions). Not surprisingly, the $EKE$ is concentrated at the surface near the front. In addition, Barth observed that the instability (and consequent loss of $APE$) leads to a flattening of isopycnals and destruction of the initial front. Durski and Allen (2005) extended these results to other configurations and demonstrated that neither realistically irregular topography nor the presence of realistic wind forcing makes a substantial difference in terms of the frontal stability or consequent nonlinear evolution.

The upwelling front provides a convenient vehicle for demonstrating the concepts introduced in the previous section. Results from an idealized frontal instability model are presented in Figure 9.2, which shows surface temperature fields at different times, as well as time series of useful diagnostics. In this model run, an upwelling-favorable alongshore wind stress blows for 6 days (gray shaded area in the $APE$ plot) and then stops. An upwelling front develops (comparable to Figure 4.2), and the sloping isopycnals (representing increasingly large horizontal density gradients over the first 6 days) fuel the growth of $APE$, which reaches a maximum shortly after the wind ceases. The front is itself baroclinically unstable; that is, $C_{PE \to KE}$ is positive (last panel), so $EKE$ begins to grow (uppermost time series) while $APE$ is being consumed. Energy is lost along the way,

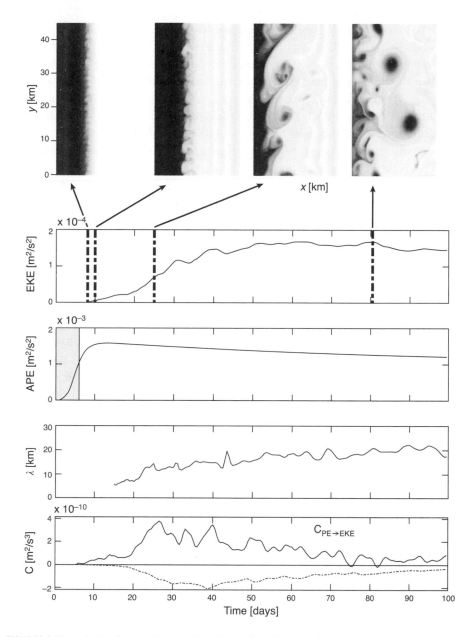

**FIGURE 9.2.** The evolution of a numerical simulation of a baroclinically unstable flow over a continental shelf. The upper four panels represent surface temperature (dark shades correspond to low temperatures, with the coastline along the western edge) at the times indicated by the heavy dash-dot lines and arrows. The next plot is spatially averaged eddy kinetic energy (*EKE*) as a function of time. Next is the available potential energy (*APE*), where the shading represents the time over which an upwelling-favorable wind stress is applied (note the scale difference!). The second frame from the bottom is a time series of the estimated dominant alongshore wavelength $\lambda$ (estimates for $t < 15$ days, when the instability is still weak, are deleted as unreliable). Finally, the lowest panel represents energy conversions. The solid line is the spatially averaged conversion from potential energy to eddy kinetic energy, and the dash-dot line is the frictional dissipation of eddy kinetic energy. These plots are based on model run 4 of Brink (2016).

however, as demonstrated by the eddy frictional-loss term (dash-dot line in the lower-most panel). While this is going on, the surface temperature field evolves. At day 6 (leftmost temperature map), there are only small perturbations in the front, and they have relatively short alongshore wavelength, well less than 10 km. As time goes on, the frontal deviations become more pronounced, have larger alongshore scale, and become more irregular (middle two maps). The estimated dominant alongshore wavelength (third time series) continues to grow until about day 60 and then does not noticeably evolve any further. By the time of maximum $EKE$ (day 80.5), the surface temperature field is quite irregular and is dominated by a couple of well-defined eddies. Notice, too, that, by this time, there is not much cold water left near the coast, because the instability has led to a flattening of isotherms (equivalent to a decrease in $APE$) and a general relaxation of the initial front. This example demonstrates how the energy conversion and other calculated quantities describe the evolving disorder that one can easily see in the surface temperature maps.

One interesting aspect of observations off Oregon is that sometimes (but not consistently, e.g., Mooers, et al., 1976; Stevenson et al., 1974) there is convincing evidence for subduction at the upwelling front, that is, offshore-moving near-surface water from inshore sinking below the surface front. For example, off Oregon, salinity often governs the upper-ocean density structure, so, to a limited extent, temperature acts like a tracer. Sometimes a tongue of warmer water passes downward and offshore beneath the salinity front (e.g., Figure 9.3A near 10 km). These subduction features, when present, are often also detectable in other fields such as chlorophyll concentration (Figure 9.3E), light transmission (Figure 9.3D), or dissolved oxygen. However, the evidence for subduction is spotty: some frontal sections show tongues, and some do not. One potential explanation for this is that the unstable jet meanders onshore and off, and that vorticity conservation in the course of meandering leads to upward and downward motions (e.g., Newton, 1978). Specifically, there will be upward motion when the jet's path has a shoreward component, and downward motion when the path tends offshore. This possibility has not been verified with coastal ocean observations, but similar dynamics have been shown to apply in Gulf Stream meanders far offshore (Bower, 1989). If this explanation is true of the upwelling front, whether one sees evidence of subduction is simply a matter of where in the unstable meandering pattern the cross-shelf section happens to lie. The problem has been investigated thoroughly using a numerical model of a comparable (but differing in detail) front (Freilich and Mahadevan, 2021). Their results support the vorticity-meandering notion, but they also show additional subduction patchiness associated with smaller-scale, $O(10$ km$)$, filamentation and instabilities.

## 9.4.   Other Wind-Driven Cases

Just as an upwelling-favorable wind stress gives rise to upward bending of isopycnals and a surface front, downwelling-favorable wind stress is expected to give rise to a front that bends downward and intersects the bottom (e.g., Allen and Newberger, 1998; section 4.6;

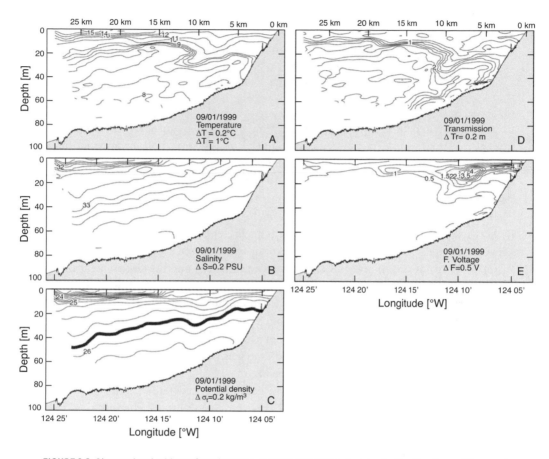

**FIGURE 9.3.** Observational evidence for subduction at an upwelling front. A cross-shelf section from off the coast of Oregon, September 1, 1999. (A) Temperature, (B) salinity, (C) density ($\sigma_\theta$), (D) light transmission (a measure of turbidity), and (E) chlorophyll fluorescence. Adapted from Austin and Barth (2002).

Figure 4.5). Such a front has little or no expression in surface temperature, for example, so it is difficult to observe in practice. Brink (2016) investigated the stability of this configuration using a numerical model driven by a wind impulse (as used with Figure 9.2) rather than using an idealized frontal initial configuration. Initially, $y$-independent symmetric instabilities and slantwise convection occur in the thick, downslope-moving bottom boundary layer, but these features are quickly overwhelmed by $y$-dependent baroclinic instabilities. Interestingly, in this case, the *EKE* is relatively evenly distributed across the shelf and is concentrated in the waters immediately above the bottom boundary layer. Wenegrat et al. (2018) investigated these transitions in more detail and also showed that the thin bottom boundary layer associated with onshore (upslope) Ekman transport is, in contrast, relatively stable to baroclinic instabilities.

In both the upwelling and the downwelling wind-driven cases, Brink (2016) found that the maximum (as a function of time) *EKE* scales as

$$EKE \approx C_0 \, APE^* \, (1+C_1 s^2)^{-1},$$    (9.4.1)

where the slope Burger number is

$$s = \frac{\alpha N}{f},$$    (9.4.2)

$\alpha$ is the uniform bottom slope, and $N$ is the initial buoyancy frequency. $APE^*$ is a scaling for the maximal wind-driven $APE$ based on initial stratification, bottom slope, latitude, wind-driven Ekman transport, and wind duration. The $O(1)$ constants $C_0$ and $C_1$ are found empirically based on many model runs, and each differs for upwelling versus downwelling. The model suggests that the eddy size (at the time of maximum $EKE$) scales as an inertial length $(\hat{u}/f)$ for small $s$ and as something more like a topographic Rhines (1977) scale $(\hat{u}/\beta_T)^{1/2}$ for large $s$ (where $\hat{u}$ is a representative velocity scale, and, in this case, the vorticity gradient $\beta_T = \alpha f/h$ is due to topography rather than to Earth's curvature). The Rhines scale marks the point where nonlinearity and the tendency for waves to propagate have equal magnitudes. For larger length scales, the Rossby number $\hat{u}/(fL)$ is small enough that Rossby waves (or topographic Rossby waves in the present context) dominate over nonlinearity, and the cascade to larger scales cannot go further. For shorter length scales, the Rossby number is larger, so eddy interaction (nonlinear) effects dominate, and scale evolution continues to generate larger eddies. Similar energy and length scalings, where eddy energy depends on $APE$ and where evolved eddy length scale depends on eddy strength, are not unusual in other coastal instability problems. It is also normal for problems involving baroclinic instability for the initial (small-amplitude) length scale to be related to an internal Rossby radius of deformation, but as the instability grows and the eddy field evolves, eddy length scale increases to the limiting values.

A continuing, purely oscillatory, spatially uniform alongshore wind forcing, starting with a resting ocean, also gives rise to baroclinic instabilities and an eddy field over the model shelf. What is interesting about this problem is that correlation as a function of separation for model $u$, $v$, or $\rho$ can be meaningfully compared with analogous observations (Figure 9.4). For example, both observations and model results are used to estimate the correlation of $u$ at one location with $u$ at another location at some distance alongshore but on the same isobath. Brink and Seo (2016) found that model alongshore currents $v$ and sea level $\zeta$ are dominated by the directly wind-forced flow and are well correlated over large alongshore distances (Figure 9.4, lower panel), while cross-shelf currents $u$ below the surface Ekman layer involve weaker wind-driven flow and so are dominated by the eddy field. Consequently, $u$ time series are correlated only over eddy-like alongshore scales of less than 10 km. Also, eddy cross-shelf current fluctuations, while weaker than alongshore currents, are considerably stronger (typically a third of the alongshore amplitude) than expected based on the coastal long-wave approximation (section 5.7). These findings are all consistent with statistical properties often observed over the shelf (e.g., Figure 9.4, upper panel; Kundu and Allen, 1976).

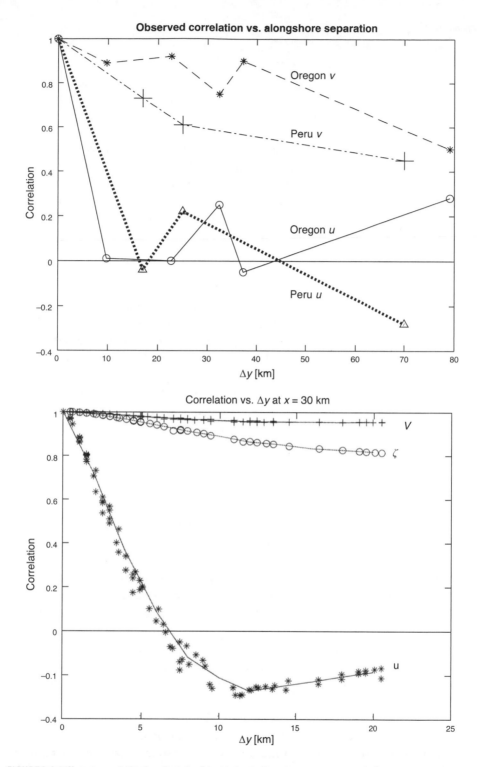

**FIGURE 9.4.** Different correlation length scales for mid-depth alongshore versus cross-shelf currents. Upper panel: observed intercorrelations of midshelf alongshore (* and + symbols) and cross-shelf (o and Δ symbols) velocity as a function of alongshore separation. Observed correlations of less than 0.3 are probably insignificant. Lower panel: intercorrelations of model alongshore velocity (+), sea level height (o), and cross-shelf velocity (*) as a function of alongshore separation. Note the different horizontal scales between the two panels. Adapted from Brink and Seo (2016).

## 9.5.    Shelf Break Front

Some of the earliest observations and models of coastal ocean frontal instability involved the shelf break front south of New England (e.g., Garvine et al., 1989; Flagg and Beardsley, 1978). This feature, which separates relatively fresh shelf waters from saltier offshore waters, consistently resides near the shelf edge, sloping upward and offshore from the bottom (typically near the 100 m isobath) to the surface (Figure 10.1). Although instabilities here have long been recognized, observing them in isolation is difficult because of the frequent presence of warm-core rings just offshore of the shelf edge (e.g., Gangopadhyay et al., 2020).

The current understanding of shelf break eddies was encapsulated by Zhang and Gawarkiewicz (2015), who examined primitive equation numerical model calculations with idealized frontal initial conditions. Energy conversions were dominated by the transfer from potential to kinetic energy (9.2.7); that is, the instability was primarily baroclinic. They found that the eddy kinetic energy is concentrated at or offshore of the initial frontal location, so the effect over the shelf proper (water shallower than about 80 m) is not pronounced. Finally, they obtained a range of equilibrated dominant eddy wavelengths of 20–70 km, depending on initial conditions, such as frontal density contrast. These results appear to be consistent with existing observations, but it remains a challenge to understand the cross-frontal fluxes associated with these eddies.

## 9.6.    Buoyancy Currents

When relatively fresh water flows out of an idealized estuary, part of it turns to the right (in the northern hemisphere) and feeds an alongshore buoyancy current (chapter 8) which tends to hug the coast far from the actual outflow. In nature, buoyancy currents span a huge range of spatial scales and of transports. Even near relatively large rivers such as the Columbia (Hickey et al., 1998), the current can be relatively thin (5–15 m thick) and very mobile in response to alongshore wind fluctuations. Thus, buoyancy currents can be difficult to track in the field, and their behavior is not as well observed as their importance would suggest. Satellite color or temperature data can sometimes be useful tools for tracking these features from space. Hetland (2017), for example, used these measurements to document the eddy-rich buoyancy current over the shelf west of the Mississippi River outflow. He observed eddy scales that ranged from tens of kilometers up to about 100 km.

Hetland then carried out a sequence of focused numerical experiments dealing with this baroclinic instability, all based on similar, idealized, alongshore-uniform initial states but with varying bottom slope, stratification, and frontal strength (among other properties). His analysis was guided by insights from relevant quasigeostrophic studies, such as those of Blumsack and Gierasch (1972). One interesting result was finding a parameter range with relatively narrow buoyant features where the currents are stable, in contrast with numerical stability studies of many other sorts of coastal contexts (e.g., Brink

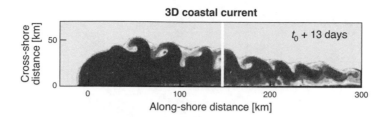

**FIGURE 9.5.** Instability of a buoyancy current. Results from a numerical model of outflow from a river (at the location where alongshore distance is zero). Surface salinity at a time more than 13 days after the start of the inflow. Darker shades represent fresher water. Adapted from Chen et al. (2019).

and Seo, 2016) that often found at least weak instability for all frontal configurations considered. Given the tremendous range of buoyancy current scales and structures, there is yet more to be done based on other starting assumptions (e.g., Chen et al., 2019; Figure 9.5) and on additional forcing types.

One important impact of buoyancy current instability is mixing of coastal with ambient waters, a major determinant of how far alongshore a current can extend before it loses its identity. Wind-driven surface Ekman transport can also lead to substantial cross-shelf exchange out of a buoyancy current (Lentz, 2004; section 8.5). It is straightforward to estimate the volume of water exchanged during a single upwelling wind event (Ekman transport times wind duration); if the initial current's volume is not large, a single wind event can substantially break down a buoyancy current. For larger scale currents, where a wind event's net transport is a smaller fraction of the initial cross-sectional area, the same wind event is relatively less effective. Thus, instability then seems less likely to be overshadowed by winds as a means of dissipating larger-scale buoyancy currents.

## 9.7.    Western Boundary Currents

Off the southeastern coast of North America, where the Gulf Stream flows along the outer edge of the continental shelf, there is a distinct front separating shelf water from warmer, much faster moving western boundary current water. Satellite temperature images (e.g., Legeckis, 1979) reveal poleward-propagating perturbations along this inshore edge of the Gulf Stream. These originate as far south as 28°N, move at about 30–50 km/day, and occur at 4-day or longer intervals. In situ measurements (e.g., Brooks and Bane, 1983) showed that the meanders are centered on subsurface shelf-edge dome-like features (of order 20 km cross shelf and 40 km alongshore) which extend down to about 300 m or deeper (Figure 9.6) and are associated with current perturbations of perhaps 0.5 m/s. What is visible in sea surface temperature is a roughly 20 m-deep streamer of Gulf Stream water that is pulled counterclockwise around the dome and can extend across much of the shelf. Typically, when the Gulf Stream turns offshore near Cape Hatteras, the meanders are severely distorted (e.g., Andres, 2021) if they have not already vanished. Energy conversion estimates comparable to (9.2.7–9.2.8) show that the features are extracting energy

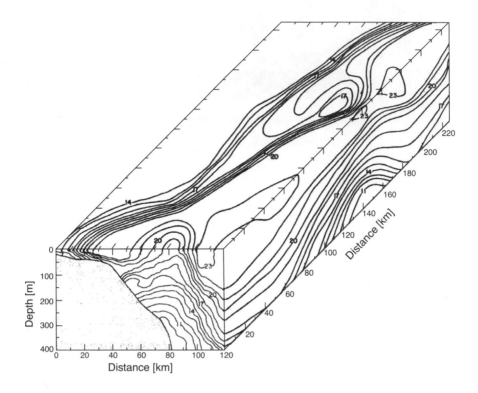

**FIGURE 9.6.** Instability along the inshore edge of the Gulf Stream. Three-dimensional plot of the observed temperature field associated with a Gulf Stream instability. The cross-shelf section shows the cold dome at the shelf edge, and the contorted Gulf Stream front shows clearly at the surface. February 11, 1979. Adapted from Brooks and Bane (1983).

from *PE* to *EKE*, while volume flux estimates demonstrate that these eddies are an important supplier of nutrients to the outer continental shelf (Lee et al., 1991). Similar features are also found inshore of the East Australian Current (Schaeffer et al., 2017) and, apparently, inshore of the Agulhas Current (Krug et al., 2017). It should be noted that the Agulhas is also home to occasional (a few times per year) distinct propagating features that can cause $O(100$ km$)$ offshore meanders and that grow via a barotropic instability transfer (9.2.8) (Elipot and Beal, 2015).

Luther and Bane (1985) carried out a linear stability analysis based on an idealized Gulf Stream and topographic configuration. The more nuanced, fully nonlinear problem was treated by Gula et al. (2015). The linearly most unstable solution closely matched the properties of the observed features in terms of spatial scale, structure, propagation rate, and the dominantly baroclinic instability origin. It is remarkable that the linear stability properties matched observations so well as compared with other cases, such as Zhang and Gawarkiewicz's (2015) shelf break instabilities, where tens of days of nonlinear evolution were required to obtain results that could be compared favorably with observations. One potential explanation of this contrast is that the Gulf Stream features,

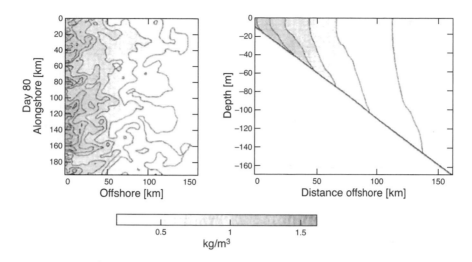

**FIGURE 9.7.** Baroclinic instability associated with surface cooling over a sloping bottom. Density anomaly at the surface (left) and as a cross-shelf section of alongshore-averaged density (right) from an idealized numerical model run driven by sustained uniform surface cooling. Darker shading represents denser water. Adapted from Pringle (2001).

moving alongshore over no more than about 7° of latitude, exist for no more than about 14 days before they lose their identity near Cape Hatteras. It thus seems possible that these features do not have enough time to evolve to a state much different structurally from the linearly unstable solutions.

## 9.8.    Wintertime Cooling

If an initially resting ocean with a sloping bottom experiences a uniform surface cooling, surface-to bottom mixing assures that the shallowest water column becomes the coldest, because the same amount of heat is extracted everywhere, so the shallowest water (least mass) cools most rapidly. To the extent that temperature governs density variations, this means that cooling creates a cross-shelf density gradient which, in turn, can become baroclinically unstable. A similar initial density distribution can also be achieved in warm, dry climates owing to evaporation, hence salinification and latent cooling (e.g., Shearman and Brink, 2010). In either case, the resulting instability decreases $APE$ by partially flattening the mean isopycnals, thus stably stratifying the water column (Figure 9.7). This basic insight, along with an appreciation for the added role of alongshore winds, date to Whitehead (1981), who demonstrated the instabilities using laboratory models. Since then, the problem has been revisited using numerical models in a range of configurations that offer a more quantitative synthesis (e.g., Chapman and Gawarkiewicz, 1997; Pringle, 2001).

Pringle (2001) considered a system, initially uniform alongshore, that has surface cooling and a sloping bottom. This becomes baroclinically unstable, and then, over a few tens of days, an eddy field develops over the shelf. The vertical structure of the eddies depends on the bottom friction, although it appears that realistic bottom friction values cause eddies to be surface intensified and thus minimally damped by the bottom stresses. In any case, the eddies transfer heat across the shelf so as to lessen the initial temperature gradient and to make the water column more nearly stable gravitationally (Figure 9.7, right panel). A downwelling-favorable alongshore wind stress weakens the instability by tilting isopycnals shoreward, thus decreasing $APE$. In contrast, wind-driven upwelling has little effect, because it carries denser nearshore water over lighter water, so vertical mixing occurs, and the density field is not greatly modified (Brink, 2017).

## 9.9.    Tidal Mixing Fronts

Strong tides give rise to large bottom stresses and thick bottom boundary layers. Simpson and Hunter (1974) provided a classic criterion for the strength of tidally driven turbulence required to overcome the tendency for surface heating to stratify the water column. Their $h/\hat{u}^3$ parameter (where $h$ is the water depth and $\hat{u}$ is a representative tidal amplitude: section 7.5) shows that complete tidal mixing is most likely in shallow water, where $\hat{u}$ is large, and becomes less likely in deeper water, where $\hat{u}$ is usually smaller via continuity. Thus, a tidal mixing front develops, roughly parallel to isobaths, which separates shallower well-mixed water from stratified waters (e.g., Figure 9.8). These fronts are common in regions with elevated tides such as Georges Bank east of North America, and around the British Isles. Very often, there is an along-front jet, and the well-mixed regions appear to sustain a high biological productivity (e.g., Cohen and Grosslein, 1987). There is substantial observational evidence for the tidal mixing fronts being unstable (e.g., Simpson and James, 1986; Badin et al. 2009).

The linear stability and eddy evolution of an idealized tidal mixing front, with no tidal currents or bottom friction, was treated by Brink (2012b), who found the front to be baroclinically unstable. If the front is very sharp, or if there is bottom friction (Brink and Cherian, 2013), then there is often a near-bottom region where initial $fQ_E$ is negative, with $Q_E$ being the Ertel vorticity:

$$Q_E = -\rho_z(f + v_x - u_y) + \rho_x v_z - \rho_y u_z. \tag{9.9.1}$$

These configurations give rise to initial symmetric instabilities and slantwise convection, but within days, the larger-scale finite amplitude baroclinic instability grows and eliminates the slantwise convection. This finding of transition between instability types occurs in at least some other settings as well (e.g., Wenegrat et al., 2018). The presence of realistically strong bottom friction allows eddies to undergo stratified spindown near the bottom, hence surface intensification; the typical eddy has no horizontal circulation near the bottom, but the swirl velocity increases monotonically toward the surface. The resulting velocity profile is sometimes referred to as "equivalent barotropic," meaning it has depth dependence, but velocity never changes sign with depth. Over time, the eddies

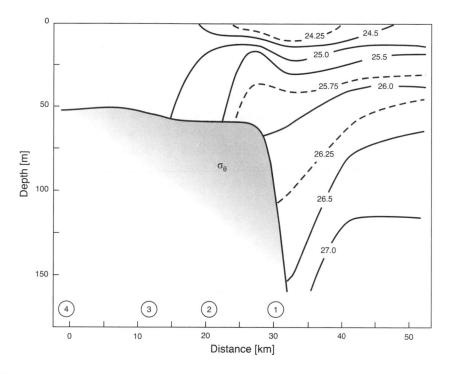

**FIGURE 9.8.** An observed tidal mixing front. A typical density ($\sigma_t$) section across the northern side of Georges Bank, July 3, 1988. Adapted from Loder et al. (1992).

evolve from a relatively short initial scale (about 10 km alongshore wavelength) to larger scales with wavelengths greater than 40 km. These relatively barotropic eddies have sizes that scale with the topographic Rhines scale $(\hat{u}/\beta_T)^{1/2}$. It may seem odd that a baroclinic instability can ultimately lead to a scale set by nearly barotropic dynamics, but the important link is the tendency for the eddies to become equivalent barotropic and for the horizontal scale to increase with time until it can grow no more.

More realistic numerical experiments impose tidal currents so that a tidal mixing front is formed by the model (as opposed to being prescribed at the outset). Despite the added time dependence and dissipation, the front remains unstable (again, primarily baroclinically), and many of the qualitative features of the idealized models carry over (Brink, 2013). Horizontal mixing associated with the eddy field is parameterized and represents a substantial transfer across the front. A simple ("NPZ") biological model is included to explore the implications of the cross-frontal transfer. Interestingly, the lateral mixing has little effect on the overall biological activity, evidently because a deep horizontal eddy flux of nutrients from stratified water to the mixed zone is balanced by near-surface lateral mixing from the mixed region into the nutrient-poor upper level in the stratified region.

## 9.10.  Surf Zone

On the innermost part of the continental shelf (water depth roughly < 5 m: section 6.4), the primary driver of mean flows and edge waves is the radiation stress associated with incoming surface gravity waves. In this region, horizontal scales of $O(100$ m$)$ and time scales of $O(100$ s$)$ are fairly common, and the Rossby number can easily be $O(100)$. Thus, Earth's rotation can be quite negligible. Under these circumstances, baroclinic instability is irrelevant, but barotropic instability can still be a factor. Indeed, Bowen and Holman (1989) developed a theory of barotropic instability in the nearshore. The necessary condition for linear instability is that $\Pi_x$ change sign for some $x$, where the potential vorticity is

$$\Pi = \frac{f + v_{0x}}{h} \approx \frac{v_{0x}}{h},$$

(9.10.1)

and $v_0(x)$ is the mean alongshore velocity. They assumed a simple jetlike flow with $|v_{0x}| = O(10^{-2}$ 1/s$)$ and found unstable wave solutions with period $O(100$ s$)$ and alongshore wavelength $O(100$ m$)$. Oltman-Shay et al. (1989) then verified the broad aspects of the theory by using nearshore observations to demonstrate the presence of a new waveform, distinct from the edge waves, having scales consistent with the model (Figure 9.9).

The nonlinear aspects of this problem were treated numerically by Slinn et al. (1998), among others. Under weakly unstable conditions, linearized solutions are found to be a good representation of observed variance. As the degree of supercriticality grows (i.e., as the basic flow becomes increasingly unstable), results become more eddy-like and chaotic. An eddy-rich field redistributes alongshore momentum, thus strongly affecting the mean flow. Given typical nearshore conditions, Slinn et al. concluded that surf zone eddies, born of barotropic instability, are likely common in nature. However, eddies observed in the surf zone, while they may be common and energetic, can also be driven efficiently by breaking incoming surface waves, so they may not require barotropic instability as a generating mechanism (e.g., Feddersen, 2014).

## 9.11.  Conclusion

Instabilities and eddies in the coastal ocean are distinctive for a number of reasons. One is that the quasi-geostrophic approximation, so useful in deeper water, fails because of the strongly sloping bottom and the vertical isotherm displacements being of the order of water depth. Further, the eddy length scales are short enough that Rossby numbers for evolved eddies are often $O(1)$, which makes the mid- and outer-shelf coastal eddy properties more like those in the oceanic submesoscale range than of the oceanic mesoscale. In addition, coastal waters are shallow enough that the effects of bottom friction, directly or indirectly, are large compared with open-ocean contexts. Because the space and time scales for coastal eddies are relatively short compared with the oceanic mesoscale, adequate synoptic field measurements are particularly demanding.

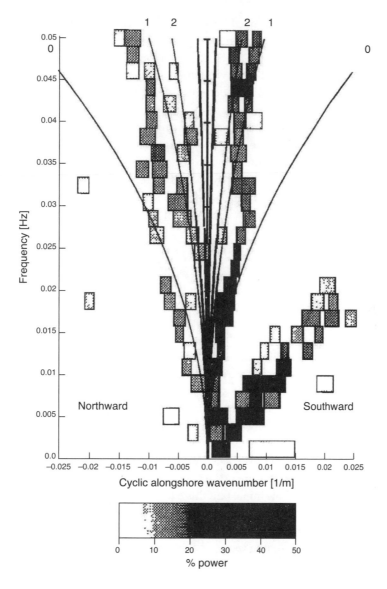

**FIGURE 9.9.** Barotropic instability in the surf zone. Observed frequency-wavenumber spectrum of alongshore velocity measured in the surf zone off Duck, North Carolina, The shading within a box darkens with greater energy content for that frequency-wavenumber pair. The dispersion curves for the first few linear edge waves (section 5.5) are shown as solid lines. The unstable wave is the lowest-frequency branch on the southward (right-hand) side. Adapted from Oltman-Shay et al. (1989).

One of the most important points about the examples in sections 9.3 to 9.10 is that there are so many coastal contexts that can credibly give rise to instabilities and an eddy field. It is hard to imagine any segment of the coastal ocean that is not affected by wind driving or (outside the tropics) by seasonal cooling. It seems likely, then, that most places in the coastal ocean are home to an eddy field, at least sporadically. This eddy field is likely not the strongest overall variability in most places (based on existing observations that are generally dominated by wind effects, coastal-trapped waves, tides, or buoyancy currents, for example), but the eddies may very well dominate the cross-shelf flow component. This insight was apparently first expressed by Kundu and Allen (1976) based on observations off Oregon.

Two indirect lines of observational evidence support this conjecture of omnipresence. One is that in those places where there are enough data, and the results are published, alongshore correlation scales for alongshore currents $v$ (associated with large-scale processes) are far larger than scales for cross-shelf currents $u$; see Kundu and Allen (1976), Winant (1983), Dever (1997), and Figure 9.4. Second, there is the difficult $fu$ term in the alongshore momentum equation. When terms in the alongshore momentum equation

$$v_t + fu = -\frac{1}{\rho_0} p_y + (stress\ terms) \tag{9.11.1}$$

are estimated using data, the $v_t$ and $p_y$ terms are often correlated with each other or with the stress terms (if the equation is depth-integrated). However, even though the $fu$ term can be directly computed from data, it is often difficult to correlate it with any other terms aside from occasionally $v_t$ or with stress terms within the turbulent boundary layers (e.g., Allen and Kundu, 1978; Hickey, 1984). People rarely, if ever, find a correlation of $fu$ with any estimate of $p_y$. Thus, $u$ measurements are often characterized as "noisy" without much further consideration. One possible interpretation, consistent with instability, of this noncorrelation is that $p_y$ is usually estimated over scales too large—usually tens of kilometers or more—to compare with the very short spatial scales inherent in $u$ and eddies.

While it appears reasonable to expect the coastal ocean at least sometimes to be eddy-rich virtually everywhere, there are not yet enough observations to be assured that this is really the case. Further, how important this eddy field might be in terms of cross-shelf transport is also unresolved. The obstacles here are evidently observational, but given ever-improving capabilities (e.g., Yoo et al., 2018; Kirincich et al., 2022), this challenge may well be met over the coming years.

# 10
# Relation to the Open Ocean

## 10.1.  Introduction: Constraints on Shelf-Ocean Coupling

In the preceding chapters, coastal ocean processes have been treated either as if the adjoining deep ocean were passive or as if its function were only to provide tidal forcing, yet the coastal and open oceans *are* coupled more closely. After all, continental shelf waters are salty (Figure 10.1), and the salt has to come from somewhere. This chapter now deals with coupling between shelf and oceanic waters. That coupling takes two general forms. One involves transporting water properties from one domain to the other, and the other involves momentum—how motions in one domain affect those in another without necessarily transporting water properties.

One overriding constraint is the Taylor-Proudman theorem. If the ocean is assumed to have steady flow, linear momentum balances and nondissipative physics, then the horizontal velocity is geostrophically balanced:

$$- fv = -\frac{1}{\rho_0} p_x,  \tag{10.1.1a}$$

$$fu = -\frac{1}{\rho_0} p_y,  \tag{10.1.1b}$$

where $(u, v, w)$ are velocity components in the $(x, y, z)$ directions, $p$ is pressure, $f$ is the constant Coriolis parameter, $\rho_0$ is a constant background density, and subscripted independent variables represent partial differentiation. The continuity equation is

$$0 = u_x + v_y + w_z.  \tag{10.1.2}$$

If $f$ is a constant, then using (10.1.1) in (10.1.2) yields $w_z = 0$, and the condition of no flow through the ocean's surface requires that $w = 0$ everywhere. Next, partial differentiation of (10.1.1) in the vertical, plus use of the hydrostatic condition

$$0 = -p_z - g\rho,  \tag{10.1.3}$$

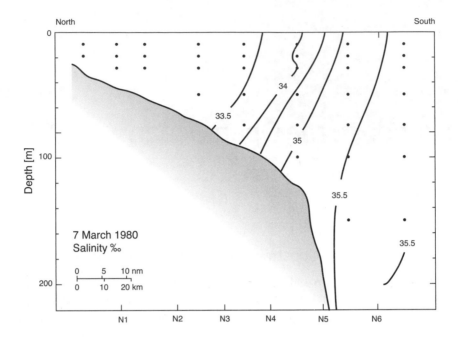

**FIGURE 10.1.** A salinity contrast between the shelf and open ocean. March (wintertime) cross-shelf section of salinity south of New England. Although shelf and oceanic waters differ in salinity by more than 2 ‰ in this case, salinity exceeds 33 ‰ across the entire section. Adapted from Beardsley et al. (1985).

(where $\rho$ is the variable component of density—which is small compared with $\rho_0$ in the ocean, and $g$ is the acceleration due to gravity) to eliminate pressure, leads to the thermal wind equations

$$\rho_x = -\frac{f\rho_0}{g}v_z, \tag{10.1.4a}$$

$$\rho_y = \frac{f\rho_0}{g}u_z. \tag{10.1.4b}$$

The equation governing density (analogous to 2.3.5d but with no mixing or time dependence) is, with $w = 0$,

$$0 = u\rho_x + v\rho_y, \tag{10.1.5}$$

or, using (10.1.4),

$$0 = -uv_z + vu_x, \tag{10.1.6a}$$

and converting to vector notation:

$$0 = \hat{k} \cdot \boldsymbol{v} \times \boldsymbol{v}_z, \tag{10.1.6b}$$

where $\hat{k}$ is the vertical unit vector. Thus, whenever $v$ is nonzero, $v$ must be parallel to $v_z$: the velocity vector cannot veer with depth (although its magnitude and sign can change with depth). Further, the bottom boundary condition

$$w = 0 = -v \cdot \nabla h \qquad\qquad \text{at } z = -h(x, y), \qquad\qquad (10.1.7)$$

where $h$ is the water depth, requires that flow at the bottom, hence (from 10.1.6) at all depths, must follow isobaths exactly. In other words, steady, linear, frictionless flow with constant $f$ cannot cross isobaths. If there is no density stratification, it is straightforward to use vorticity conservation (e.g., 8.4.2a) to show that

$$0 = v \cdot \nabla \left( \frac{f}{h} \right), \qquad\qquad (10.1.8)$$

that is, that steady, barotropic, adiabatic, linear (i.e., geostrophic) flow must follow "geostrophic contours," that is, lines of constant $f/h$. Because $w \neq 0$ when $\nabla f \neq 0$, the stratified result (10.1.6) actually holds only approximately, over scales much smaller than Earth's radius. Continental shelves and slopes indeed obey this limited-scale criterion.

Results (10.1.6–10.1.8) place strong constraints on the strength of cross-isobath flow relative to along-isobath flow. Cross-isobath transports can clearly be enhanced if the Taylor-Proudman constraint is weakened. Thus, the starting assumptions of the preceding derivation must be reconsidered if this chapter is not to end here. Letting $f$ vary with latitude does weaken this constraint slightly, but the path of $f/h$ contours over most continental slopes is so nearly along isobaths that it does not provide a terribly effective enabler in this context. Considering the other assumptions used, it appears that the major enablers of cross-isobath velocity are time dependence, dissipation, and nonlinear effects.

However, the strength of the cross-shelf flow is not the only consideration here; the spatial scales of interaction are also important. Physically, even a weak cross-shelf flow can be effective if it operates over a large enough alongshore scale. When the continuity equation (10.1.2) is scaled, it leads to the conclusion that

$$L^y = O\left( \frac{\hat{v}}{\hat{u}} L^x \right), \qquad\qquad (10.1.9)$$

where $(L^x, L^y)$ are typical cross- and along-isobath (cross-shelf and alongshore, in the coastal context) scales and $(\hat{u}, \hat{v})$ are corresponding velocity component scales. Thus, even if the cross-shelf velocity is weak, it can still be effective when integrated over alongshore scales much greater than $L^x$, the shelf width. Similarly, even a very strong cross-shelf flow may be ineffective on a regional scale if it is not sufficiently widespread or sustained. Thus, the effectiveness of a given mechanism depends on its local strength, as well as on its spatial scale and temporal persistence.

The following discussion is built around the three broad enablers of cross-isobath velocities. This chapter thus embraces a wide assortment of topics, and some of them, such as tides, have already been treated elsewhere in this volume. Given the diversity of processes involved, most are touched on only briefly here, with the hope that the references and summaries provide a useful introduction. The key to evaluating these various shelf-ocean couplings is always the integration of the transfer over space and time, rather than simply the instantaneous or localized velocity.

## 10.2.   Time-Dependent Processes

### Tides and Higher Frequencies

Ocean tides drive continental shelf tides (chapter 7) and so provide a prime example of momentum transfer across the continental margins. One reason that this transfer is so effective, of course, is that tidal frequencies are typically of the same order as $f$ (or substantially larger near the equator), assuring a clear violation of the steady-flow assumption in section 10.1. In particular, Poincaré wave resonances can expedite very strong momentum transfers. In addition, internal tides (section 7.6) can propagate onto the shelf, in some cases from far offshore, and as they pass into shallow water they create turbulence, contributing to shelf mixing. Even more effective is the propagation of high-frequency gravity waves onto the shelf (section 6.3) and into the surf zone, where they ultimately drive both steady and unsteady currents.

### Equatorial Processes

Equatorial waves also interact effectively with the coastal ocean (e.g., Moore, 1968; Anderson and Rowlands, 1976). For each vertical baroclinic mode (see section 5.4), several classes of waves (Figure 10.2) exist that are trapped near the equator far from shore (e.g., Clarke, 2008): (1) equatorial gravity waves, which exist at higher frequencies and are analogous to Poincaré waves at higher latitudes; (2) a single Yanai (or mixed Rossby-gravity) wave, which always has eastward group velocity but whose phase velocity changes sign with zonal wavenumber; (3) a single equatorial Kelvin wave that propagates nondispersively eastward; and (4) Rossby waves, which exist only at lower frequencies and have westward phase velocity even though the group velocity can go in either direction.

When an eastward-propagating equatorial wave encounters a bounding coastline (such as that of Ecuador in South America), the wave can sometimes reflect, either fully or partially, as a westward-propagating equatorial mode if such modes exist at the incoming wave's frequency, that is, if gravity or Rossby waves are available at the given frequency. In the case of partial reflection or if appropriate westward-propagating waves are absent, the impinging wave can excite coastal waves (essentially coastal internal Kelvin waves at very low latitudes) that propagate in both poleward directions alongshore away from the equator (e.g., Figure 10.3). The various frequency bands for reflection or transmission are different for each vertical mode because wave properties depend on the given vertical mode's long gravity wave phase speed (eigenvalue) $c_n$. However, to the extent that the system is linear, each vertical mode should behave independently so long as the bottom is flat. One observed example of eastward-traveling, first baroclinic mode waves (Enfield et al., 1987) involves a Yanai wave which is excited by zonal winds in the

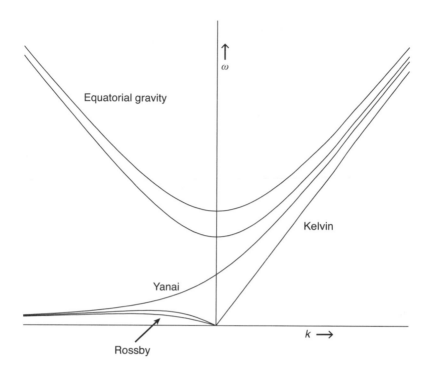

**FIGURE 10.2.** The variety of waves propagating along the equator. A schematic equatorial wave dispersion diagram (frequency $\omega$ as a function of east-west wavenumber $k$) for a single baroclinic mode. Only the two highest-frequency Rossby waves are shown, and only the two lowest-frequency equatorial gravity waves are shown. There are actually infinite sets of these two classes, with higher modes having more complex north-south structures.

equatorial Pacific. When this wave encounters the South American coast, it excites poleward-propagating coastal-trapped waves which can dominate shelf currents out to at least 15°S (e.g., Smith, 1978). This process is particularly obvious for periods in the range of about 1–2 weeks, the frequency window over which only eastward-propagating equatorial waves (Yanai and Kelvin) exist for the first baroclinic mode. Another example involves the impingement of an eastward-propagating equatorial Kelvin wave in the Atlantic at intraseasonal (90–120 day) time scales (Imbol Koungue and Brandt, 2021). Once again, the equatorial wave excites coastal disturbances that propagate poleward along the west coast of Africa.

So far, only eastward-propagating equatorial waves have been mentioned. What about the opposite (western) side of the ocean? In this case, long coastal-trapped waves propagate only equatorward, so they cannot carry energy away poleward along the coast. Further, eastward-propagating waves are available at all frequencies, so equatorial wave reflection is always possible, compounding the inability of equatorial waves to generate poleward propagating coastal-trapped waves.

A second important process operates on a longer, interannual time scale: El Niño (e.g., Clarke, 2008). In this case, changes in winds over the equator excite an eastward-propagating

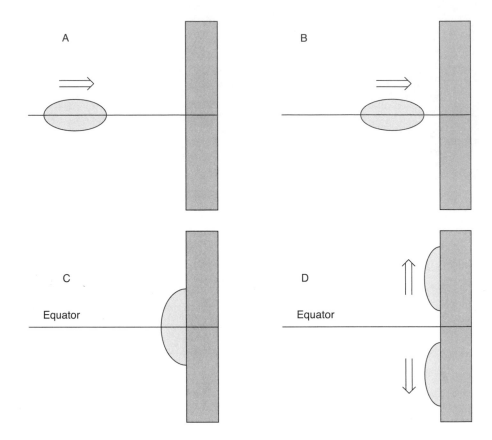

**FIGURE 10.3.** Schematic of an eastward-propagating equatorial Kelvin wave encountering an eastern boundary in a frequency range where no energy is reflected along the equator. Time advances from panel A to panel D. The lightly shaded region represents the surface elevation or thermocline depression associated with the Kelvin wave which approaches the coast (panels A–C) and then forms two coastal Kelvin waves (one in each hemisphere) propagating poleward (panel D). Motivated by the model of Anderson and Rowlands (1976).

equatorial Kelvin wave which deepens the thermocline. When this wave encounters the coast, it again generates poleward-propagating disturbances that at such long time scales do not behave as simply as most of the waves described in chapter 5. These interannual-period waves are strongly affected by bottom friction, so associated alongshore currents are greatly weakened over the shelf and relatively strengthened over the slope (Clarke and Van Gorder, 1994). Away from the equator, coastal energy can radiate westward into the open ocean as long Rossby waves (when the frequency is low enough to allow baroclinic Rossby waves in the open ocean). The net effect is that the heavily modified interannual waves propagate poleward alongshore at speeds of around 0.3 m/s, consistent with sea level observations, but far slower than the 2–5 m/s typical of higher-frequency coastal-trapped waves (section 5.6) off the west coast of the Americas. In addition, the El Niño equatorial wave that depresses the thermocline, if it has a large enough amplitude, carries some upper-ocean water with it. When such a large-amplitude wave encounters the coast, it carries water into the coastal ocean from offshore. So, sufficiently energetic equatorial waves can transport water, as well as momentum, into the coastal ocean.

Because $f \to 0$ at the equator, both Yanai waves and El Niño propagation represent "high-frequency" waves (in the sense of wave frequency greater than $f$) right at the equator. But, far from the equator, the two examples are radically different. The 1–2-week disturbances propagate as low-mode coastal-trapped waves, have no tendency to radiate energy to the open ocean, and should conform to classical theory, as outlined in section 5.6. The multiyear waves, in contrast, have radically different cross-shelf structure and alongshore propagation, including a clear tendency to radiate energy into westward-propagating oceanic Rossby waves.

## Midlatitudes

On the western side of the ocean, for example, along the eastern coast of North America, long coastal-trapped waves propagate equatorward. A simplified (but not simple) barotropic model (Wise et al., 2020) of this regime demonstrates that at low enough frequencies, coastal energy can spread eastward offshore. Further, at these low frequencies, wave structure over the shelf and slope is contorted by friction and by planetary vorticity gradients. The resulting wave structure loosely resembles a traditional shelf wave (e.g., section 5.5) over the sloping topography but is coupled to an evanescent, short Rossby wave representing a response in the ocean basin. All told, the very-low-frequency waves are almost a distinct entity compared with higher-frequency, weather-band, coastal-trapped waves. Along the eastern side of the ocean, the stratified low-frequency shelf response can (depending on frequency) radiate some energy westward offshore as a long internal Rossby wave, and the shelf-ocean coupling includes the development of an undercurrent over the slope in a stratified ocean (Samelson, 2017).

Shelf currents can respond to fluctuating oceanic flow at midlatitudes under limited conditions. The linear, stratified problem with bottom friction was treated by Huthnance (2004), who found that forcing imposed from the deep ocean leads to a substantial current amplitude on the shelf under two circumstances. One is the presence of a resonance with a coastal-trapped wave mode, which requires a match of both frequency and alongshore wavenumber. This resonance appears to be improbable, since low-mode, long, weakly damped, coastal-trapped waves tend to occur only for $(\omega, l)$ pairs that correspond to alongshore propagation speeds of $O(1 \text{ m/s})$ or greater. The ocean mesoscale, on the other hand, is typified by eddy translation speeds about an order of magnitude slower (e.g., Chelton et al., 2011). The failure of low-frequency oceanic waves to propagate onto the shelf can also be illustrated more directly. There is substantial observational evidence that when Gulf Stream meanders encounter the continental rise's sloping bottom, they excite topographic Rossby waves well offshore of the shelf edge (e.g., Pickart, 1995). When these waves encounter the steeper continental slope topography, they are evidently deflected alongshore away from crossing onto the shelf proper (Louis and Smith, 1982).

The second condition that allows oceanic fluctuations to drive shelf variability occurs when the alongshore length scale becomes very large relative to shelf-slope width. To illustrate this possibility, it is useful to simplify to the barotropic, frictionless coastal long-wave problem (chapter 5, eqn. 5.7.3 with $\sigma_F = 0$) to

$$0 = (hp_{xt})_x + fh_x p_y, \tag{10.2.1}$$

where the rigid-lid approximation has also been made. If the forcing at the offshore boundary goes as $\exp[i(ly - \omega t)]$, then the continental shelf pressure is $p = P(x)\exp[i(ly - \omega t)]$, and scaling (10.2.1) yields a cross-shelf length scale of

$$L_\omega^x = \left(\left|\frac{\omega H}{f\alpha l}\right|\right)^{1/2}, \tag{10.2.2}$$

where $H$ is a typical water depth, and $\alpha$ is the bottom slope. As might be expected based on section 10.1, this length scale goes to zero (no cross-shelf transfer) as $\omega \to 0$ or as $f\alpha$ becomes large. In contrast, cross-isobath transfer becomes larger as the alongshore scale increases ($l$ decreases), meaning that fluctuations on the shelf can be substantial if there is a very large alongshore distance over which to operate. Indeed, the alongshore scales of major coastal regions, such as the Patagonian shelf or the U.S. west coast, are typically in the range of 1000–2000 km, so substantial cross-shelf extension on this large a scale (i.e., this small an $l$) appears viable. In the extreme, two-dimensional limit $l \to 0$, there is no longer any geostrophic cross-shelf flow, the cross-shelf scale $L_\omega^x$ becomes infinite, and topography ceases to inhibit the remaining, ageostrophic cross-shelf flow. The Taylor-Proudman theorem (section 10.1) does not apply in this case because of the time dependence $\omega \neq 0$: when $\omega$ approaches zero, $L_\omega^x$ also approaches zero (there is no cross-isobath flow), just as expected.

## 10.3. Frictional Effects

The steady, frictional linear problem is most easily appreciated if stratification is neglected, and the coastal long-wave assumption is applied. Being a steady problem, no resonances are possible. The governing shelf-slope equation, from (5.7.3) with no wind forcing, becomes

$$0 = (\sigma_F p_x)_x + fh_x p_y, \tag{10.3.1}$$

where $\sigma_F$ is a bottom resistance coefficient. If forcing offshore goes as $\exp(ily)$ and pressure as $p = P(x)\exp(ily)$, then scaling (10.3.1) leads to a frictional cross-shelf length scale of

$$L_F^x = \left(\left|\frac{\sigma_F}{f\alpha l}\right|\right)^{1/2} \tag{10.3.2}$$

Substantial intrusion onto the shelf is thus possible with strong bottom friction or (once again) for large alongshore scales, but it is inhibited for large $|f\alpha|$. This steady, linear, dissipative problem was also formulated with density stratification by Kelly and Chapman (1988), and their conclusions with regard to scales remain consistent with (10.3.2).

Once stratification is included, the possibility arises that buoyancy arrest in the bottom boundary layer (section 3.7) could occur and effectively make $\sigma_F = 0$, thus preventing cross-shelf extension. Brink (2012a) explored this problem and found that, indeed, buoyancy arrest does enter the steady problem but only over an alongshore adjustment scale (e.g., 3.7.13) that increases as stratification weakens or onshore volume flux increases. Dependencies on $\alpha$ and $f$ are more nuanced than in (10.2.2) and (10.3.2) and depend somewhat on inflow configuration. For realistic parameters, the buoyancy arrest alongshore adjustment scales are typically several hundreds of kilometers or more (e.g., Table 3.1). Thus, unless topography and offshore forcing have alongshore scales of $O(1000\ \mathrm{km})$ or more, complete buoyancy arrest is not expected on the shelf/slope. While basin-scale ocean circulation has scales this large, irregular bottom topography introduces alongshore scales much shorter than typical arrest scales, so topographic irregularity is likely to prevent substantial buoyancy arrest, even on the longest time scales. Thus, it seems probable that models, such as Kelly and Chapman's, that neglect buoyancy arrest remain valid and that bottom friction can enable cross-shelf flow on regional scales.

Observationally, there is convincing evidence that on time scales of several years and longer, low-frequency sea level signals pass onto the shelf along the western side of the North Atlantic (Hong et al., 2000). The incoming ocean signal, consisting of baroclinic Rossby waves, is apparently generated by wind-stress curl distributed across the ocean basin. The Rossby waves are evidently not dominated by bottom friction in the deep ocean, but it seems very likely on these time scales that frictional effects are a substantial factor in the transmission to shallower water.

The scale dependence of cross-isobath flow leads to an important caution about relying on two-dimensional (cross-shelf, vertical) ocean models (chapter 4) too literally. These simple models themselves, of course, provide their own cautions by means of unrealistic aspects of solutions far from shore or at long time scales. Because these simplified models assume an infinite alongshore scale ($l = 0$), there appears to be nothing to inhibit transports across isobaths; that is, $L_\omega^x$ and $L_F^x$ both become infinite. But, in reality, alongshore scales are always finite, so cross-isobath intrusion is attenuated for both steady and unsteady linear conditions (10.2.2 and 10.3.2). In summary, the important point is that linear models show that the extension of oceanic forcing across the continental margin increases as the alongshore scale increases.

## 10.4.   Nonlinearities

One notion of nonlinearity is that a flow phenomenon advects momentum, mass, or vorticity to a substantial extent. Oceanographers often measure this in terms of the Rossby number: the ratio of advective acceleration, for example, in (2.3.5b), relative to the Coriolis force:

$$Ro = \frac{\hat{v}}{fL},$$  \hfill (10.4.1)

where $\hat{v}$ is a representative velocity scale, and $L$ is a typical horizontal length scale. The Rossby number can also be thought of as the ratio of relative vorticity $\hat{v}/L$ to planetary vorticity $f$. The previously discussed El Niño example demonstrates that adding nonlinearity to a process (in this case, the equatorial Kelvin wave) can enable transfer of oceanic water onto the continental shelf. Another essentially nonlinear process, instability (chapter 9), can also facilitate shelf-ocean exchange by means of a turbulent eddy transport. Especially relevant instabilities include those at the New England shelf break (section 9.5) and at the inshore edge of the Gulf Stream (section 9.7). Observational quantification of the associated eddy transports is difficult owing to the sampling requirements for estimating a statistically significant flux (e.g., Gawarkiewicz et al., 2004). The following paragraphs touch on a number of other examples of nonlinear phenomena that are relevant to shelf-ocean exchanges.

A particularly striking example of inertia locally overwhelming the Taylor-Proudman theorem occurs in the western North Pacific between Taiwan and Japan, where the shelf break changes orientation downstream from north-south to east-west. Often, the northward-moving Kuroshio Current simply turns eastward with the topography and does not cross isobaths. However, the Kuroshio occasionally remains briefly on a northward course and so crosses temporarily onto the shelf before turning eastward again (e.g., Velez-Belchi et al., 2013). These localized, transient intrusions appear to be correlated with fluctuations in the current's transport. While these events are well documented, there is much that has yet to be understood. For example, one might expect such an incursion to excite some form of coastal-trapped wave and thus have an expression at remote alongshore locations. Further, it is not at all certain to what extent these intrusions leave a permanent signature on shelf waters; for example, do they represent a net transport of salt or nutrients onto the shelf? Or does the Kuroshio water simply slosh onto the shelf and then back offshore without leaving a lasting signature? These water-mass questions are particularly difficult to address with traditional ship-based surveys because answers would require repeated field studies of events that are difficult to predict with the sort of lead time required for ship scheduling.

Another phenomenon, presumably also related to the presence of a western boundary current, occurs near Cape Hatteras, inshore of where the Gulf Stream turns from flowing parallel to the shelf break away toward the open ocean (Savidge and Savidge, 2014). South of the cape, mean flow over the shelf is poleward, in the same direction as the adjoining western boundary current. In contrast, the mean shelf flow is equatorward north of Cape Hatteras, in the Middle Atlantic Bight, so there is a sustained convergence, marked by a persistent front across the shelf near the cape. Water-mass budgets thus imply a sporadic but sustained $O(0.3 \times 10^6 \text{ m}^3/\text{s})$ offshore flux of relatively fresh shelf waters (Todd, 2020). Indeed, hints of this flux, in terms of occasional freshwater parcels in the northern wall of the separated Gulf Stream, have been observed for many years (e.g., Ford et al., 1952). The occasional, small-scale bursts that appear to make up the offshore transport seem not to represent a single sort of process. For instance, Savidge and Austin (2007) observed a near-surface, 10–20 km wide offshore-directed jet moving at >0.5 m/s on the outer edge of the shelf. In contrast, measurements by Han et al. (2021) detected a 5-day wintertime "cascading" (e.g., Shapiro et al., 2003) event in which dense shelf water spills off the shelf and down along the slope. For a typical sinking parcel, the speed and scale again appear to be around 0.5 m/s and 20 km, respectively. These contrasting offshore pathways share a couple of properties: they are both intermittent and appear to have a Rossby number of $O(0.3)$, indicating a substantial role for nonlinearity. A comparable cross-shelf frontal feature exists inshore of the Brazil Current off South America (Piola et al., 2008), so the Hatteras convergence and export may well represent a widely significant phenomenon.

Substantial offshore transports can also occur when an alongshore current over the shelf or slope encounters an abrupt change in coastal orientation. For example, consider the case where, facing downstream in an alongshore flow, the coastline and isobaths to the right of the current take an abrupt 90° turn to the right in the northern hemisphere. In perfectly linear, inviscid conditions, the flow would follow the isobaths and turn to the right as well (e.g., Crépon et al., 1984). But if the flow has substantial inertia (in the sense of a finite Rossby number), there is a tendency for the flow to overshoot the turn and depart from the isobaths into deeper water. Indeed, Nof et al. (2002) considered a very idealized version of this problem and showed that a portion of the incoming current turns to follow isobaths, while the remainder moves forward to create a chain of eddies that propagate into the open ocean. The closely related estuarine outflow problem treated by Nof and Pichevin (2001) would lead one to expect that, under realistic circumstances, the proportion going offshore versus turning alongshore would depend on the strength of the incoming flow. Observations and a realistic numerical model presented by Jithin and Francis (2021) make a convincing case that this is an important source of eddies in the Bay of Bengal and that the eddy generation versus alongshore flow tradeoff is modulated by the strength of the incoming coastal flow.

Western boundary currents such as the Gulf Stream or Kuroshio, once separated from the shelf edge, often shed energetic 150–300 km eddies (called warm-core rings in the North Atlantic) that drift poleward and westward. These detached features frequently encounter the shelf edge south of New England (e.g., Figure 10.4), and are well documented, especially in terms of occurrence at the shelf edge (e.g., Gangopadhyay et al., 2020). Numerical model studies (e.g., Cherian and Brink, 2016) predict that unless the shelf break is deeper than the $O(1 \text{ km})$ scale depth of the ring, the onshore motion of the eddy's core is halted at the shelf edge, as is always observed. Observations show that warm-core rings encountering the shelf edge occasionally entrain filaments of shelf water outward (transport of order $5 \times 10^4 \text{ m}^3/\text{s}$) or temporarily inject small bodies of warm, salty ring water onto the outer shelf. Model studies also consistently demonstrate that an alongshore extrusion of ring water should extend westward (for the New England case) along the shelf edge, and this tongue can itself become unstable, shedding small eddies (Shi and Nof, 1993; Zhang and Gawarkiewicz, 2015; Cherian and Brink, 2016, 2018). Lee and Brink (2010) were fortunate enough to be on hand to make detailed measurements of a

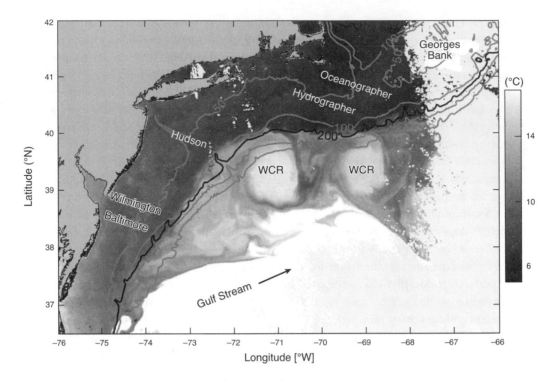

**FIGURE 10.4.** Open-ocean eddies, that is, warm-core rings (labeled WCR) encountering the shelf edge. Satellite sea surface temperature image from April 3, 2012. Darker shades represent colder water. The warm Gulf Stream is denoted by a black arrow. The 50, 100, 200, 1000, and 2000 m isobaths are shown, with the 200 m contour in black. Labels such as Baltimore and Hudson are the names of shelf-edge canyons. The white region with ragged edges represents clouds. Adapted from Li et al. (2021).

ring incursion onto the outer shelf and of the consequent small (order 25 km in diameter) eddy that formed there. In addition, it appears that near-bottom intrusions of salty off-shore water are not unusual even several tens of kilometers inshore of the shelf edge (e.g., Mountain et al., 1989), and these intrusions may well be associated with rings (Chen et al., 2022). However, the overriding remaining question about the impact of these salty intrusions is, how much does mixing allow them to make a lasting impact on shelf-water properties?

Interaction of oceanic eddies with shelf waters is not unique to the Middle Atlantic Bight. Clearly analogous features are found elsewhere, including in the East Australian Current system (e.g., Malan et al., 2020). Of course, western boundary current systems are not the only places with eddy variability in the adjoining ocean. Satellite remote sensing (e.g., Chaigneau et al., 2009) has demonstrated the existence of significant eddy fields along the eastern boundaries, in both hemispheres, of both the Pacific and Atlantic Oceans. Except for in the southeastern Atlantic, none of these regions have any connection to a western boundary current. Within these and other eddy fields, it is then not unusual to find narrow (a few tens of kilometers), rapidly moving surface filaments of cold, recently upwelled water extending up to hundreds of kilometers offshore (e.g., Strub et al., 1991; Fischer et al., 2002). These narrow features have been shown to be the expression of upwelling waters being drawn offshore around the edges of offshore eddies. The filaments typically narrow down and eventually vanish offshore. This extinction appears to be due to the cooler, denser water sinking as it moves offshore (e.g., Kadko et al., 1991), thus accounting for its disappearance at the surface.

Submarine canyons and valleys can be found on most of the world's continental shelves and, for a number of reasons, appear to be important locations for cross-shelf exchanges (Allen and Durrieu de Madron, 2009). These indentations can take the form of dramatic incisions, hundreds of meters deep, at the shelf edge (e.g., Figure 10.4); shallow depressions traversing the shelf proper; or sometimes complex, deep structures (such as Monterey Canyon off California) that extend from the slope almost up to the land. One defining aspect of canyons is their short alongshore scale (i.e., canyon width) of typically 10–20 km. Consequently, the local Rossby number (10.4.1) is relatively large, signifying a potential disruption of geostrophic alongshore flow and enabling ageostrophic processes. For example, it is occasionally observed that there is a bias toward onshore flow within canyons or valleys, e.g., Lentz et al. (2014) in the Hudson Shelf Valley south of New York City. Zhang and Lentz (2017) provided a lucid explanation for this asymmetry. Specifically, alongshore flow in the direction opposite to long coastal-trapped wave propagation tends to set up a standing wave near the valley, while ambient flow in the opposite sense merely adjusts to follow isobaths without setting up a standing wave. The net result is that the asymmetry of the response to alongshore flow leads to a bias toward onshore flow within the canyon. A very different canyon effect occurs when shelf waters become dense relative to ambient waters and then cascade down along the continental slope. Canyons are then preferred sites for these gravity currents (e.g., Wåhlin, 2002), which, in turn, are a key link in oceanic deep-water-mass formation, for example. A third important canyon phenomenon is that they are often sites of augmented internal tide or internal wave activity, leading to greatly enhanced turbulent vertical mixing (e.g., Waterhouse et al., 2017), which can, in turn, lead to horizontal mixing via shear dispersion (e.g., Young et al., 1982). Submarine canyons and valleys present a complex range of possibilities (involving nonlinearities, time dependence, and/or dissipation) for enhanced cross-shelf exchange. It remains a challenge to encapsulate these localized effects into larger-scale syntheses and thus understand the extent to which "special" locations such as these contribute to regional-scale net shelf-ocean exchanges.

## 10.5.   Sea Level and Climate Change

One might naively think that, from a coastal oceanographer's viewpoint, sea level change due to climate variation is a simple problem. It is known with some confidence that global mean sea level is currently rising at a rate of about 3 mm/year, that about 44% of that change is due to melting of ice residing on land, and that about 42% is associated with seawater expanding as it warms (WCRP Global Sea Level Budget Group, 2018). The IPCC estimates that by the end of the twenty-first century, sea level will likely rise by $O(1 \text{ m})$ (Pörtner et al, 2019). Simplistically, a globally uniform increase would mean that the sea level change imposed by the open ocean onto the shelf would penetrate readily onto the shelf, given the linear reasoning of sections 10.2 and 10.3. Of course, this is far too simple a notion.

First, sea level change in the open ocean is far from spatially uniform. There is basin-scale variability in the observed rates of change, and there are even limited regions, such as in the northeast Pacific, where open-ocean sea level is falling slightly (e.g., Stammer et al., 2013). Sometimes, these variations have relatively large alongshore scales, but there are places (like off the eastern coast of Australia) where alongshore variations can be relatively abrupt—over scales of perhaps a couple hundred kilometers.

As climate changes, winds and buoyancy forcing over the open ocean are expected to change. As these forcings evolve, the ocean circulation itself is expected to respond. This, in turn, ought to give rise to spatially variable sea level adjustments (e.g., Ponte et al., 2020). An appreciation of the magnitude of this effect might be had from the Gulf Stream, which is associated with an $O(1 \text{ m})$ cross-stream difference in free-surface height. Since this is believed to be ultimately largely wind driven (and since Gulf Stream variations are known to affect coastal sea level, e.g., Ezer et al., 2013), a radical change in open-ocean circulation would presumably allow regional sea level changes of no more than that magnitude. Similarly, changes in alongshore winds (driving alongshore currents over the shelf: chapters 4 and 5) and in surface waves and cross-shore winds (driving set-up: chapter 6) over the shelf could also lead to sea level changes (annually averaged, say) of up to a few tenths of a meter. Comparable changes in sea level might also be associated with changes in atmospheric pressure, hence "inverse barometer" effects. In addition, changes in the hydrological cycle would, in turn, affect river outflows and where water accumulates on continents, both of which could have substantial effects on sea level within estuaries and accompanying effects as buoyancy currents flow alongshore over the shelf.

The actual regional change in sea level relative to land is affected by changes in the land elevation itself. For example, it is well known that the land level in Scandinavian locales is rising at a rate of several millimeters per year (e.g., Stammer et al., 2013) owing to the rebound of the solid Earth once the glacial ice sheets, which had weighed down the land, vanish. Just as past ice sheets are associated with current glacial rebound, ongoing ice melt in Greenland, for example, can be expected to cause land levels there to rebound similarly over time scales of many centuries. In other places land is sinking locally, thus enhancing climate-induced sea level change. Notable examples include Tokyo and Shanghai, where groundwater withdrawal led to land subsidence of a few meters during the twentieth century (Nicholls, 2007).

Finally, existing ice sheets are massive enough that one needs to account for the gravitational pull of today's ice sheets versus the effect of distributing that mass (as water) relatively uniformly once melting occurs (e.g., Mitrovica et al., 2001). For the U.S. west coast, the effect of these gravitational changes due to the melting of distant ice sheets is comparable to the rate of sea level change associated with the increase in seawater volume due to that melting (National Research Council, 2012). Presumably, this gravitational effect is a substantial contributor to the global open-ocean spatial variability in sea level change noted previously.

This listing of physical processes involved in climate-driven sea level change is just the beginning. Erosion, land usage, and wetland status further affect local relative sea level and demonstrate the importance of anthropogenic factors in making local and regional assessments. In the context of this chapter, the important point is that coastal sea level rise is part of a global-scale problem and that local, shelf-scale physical oceano-graphic processes represent only a contributor, and maybe not a dominant one, to the assortment of factors involved. Indeed, many of the relevant processes (such as glacial rebound and gravitational effects) issues extend well beyond physical oceanography. For further information about the multiple processes involved, the reader might consult Caze-nave et al. (2017), Horton et al. (2018), Kopp et al. (2015) and Ponte et al. (2020).

## 10.6.  Conclusion

Knowledge of shelf-ocean interactions remains incomplete. For example, in the very well studied Middle Atlantic Bight, the shelf heat and salt inventories are well described, and a budget (Lentz, 2010) does a good job constraining the various net transports, but the detailed balances remain uncertain. For example, the actual mechanism for the required cross-shelf eddy salt flux is not firmly established. Another puzzle involves the coastal upwelling region off the U.S. west coast: the surface offshore Ekman transport is obvi-ous, but the offshore/alongshore origin of the upwelled water that compensates for the Ekman transport is yet poorly constrained. Observations (e.g., Huyer et al., 1987) and numerical models (e.g., Pringle and Dever, 2009) provide some valuable insight on this point, but a predictive synthesis remains elusive. For yet another ongoing puzzle, in the South Atlantic Bight (the U.S. east coast south of Cape Hatteras), mean alongshore flow[1] over the shelf is northward (e.g., Lee et al., 1984), in the same sense as the Gulf Stream flow; is this a coincidence?

Obtaining unambiguous, statistically significant estimates of shelf-ocean transfers remains difficult, even though observational tools are improving dramatically. For example, cross-shelf velocity (at least on time scales longer than a few days) is almost invariably weak relative to alongshore flows. This means that small errors in defining the true alongshore direction can lead to major errors in cross-shelf flux estimates. This ques-tion is complicated by the reality that bathymetric contours are generally not straight but contorted on a range of scales. Thus, in trying to determine "alongshore" by looking at a map, one needs to use considerable judgment to know over what scale to average the topography. In addition, cross-shelf flows are often episodic in time and frequently have

---

1 Mean conditions are particularly hard to deal with observationally, partly because of the need to obtain enough data to compute an actual mean field with confidence and partly because there is only one mean field. In contrast, in studying fluctuations, observations provide many events (i.e., realizations), so correlations or other statistical measures of relatedness are valuable tools.

short, $O(10 \text{ km})$ alongshore scales (sections 9.4, 9.11, 10.4), thus complicating measurements of net onshore/offshore fluxes.

These challenges present opportunities for new concepts and approaches. Refined dynamical insights will help elucidate how information and materials can pass between the shelf and the deep ocean. However, the ultimate need is to make conclusive observations that can clarify pathways and test model results. Committed, long-term measurement systems are required for these tests and are certain to bring additional, unexpected insights as well. Introduced or naturally occurring tracers, by their nature, integrate information and may lead to fresh insights as to where and how rapidly transports occur. Beyond these foreseeable approaches, the coming decades will likely introduce new and useful tools that are hard to even imagine as of this writing.

# 11
# Perspective

## 11.1. Introduction

The preceding chapters have concentrated on dynamical processes over the continental shelf, and on results dating from the mid-1960s to the present, over which time much has changed. The community's demographics have evolved from the stage during which women scientists were a rarity, to today's less imbalanced conditions. The funding climate has certainly changed with regard to competitiveness and direction. In terms of impact on our physical understanding, the most prominent changes involve advances in observational techniques, theoretical concepts, and the power of numerical models. The following sections reflect on these many advances and look to the future of our science.

## 11.2. Observational Capabilities

In the 1960s, water properties were only beginning to be measured by continuous electronic profilers (now known as CTDs: conductivity-temperature-depth). Indeed, in the early 1980s I took part in a cruise where we reverted to traditional bottle/reversing thermometer measurements because of technical difficulties with the newer instruments. At present, CTDs are easier to use, more flexible, and vastly more reliable. Contemporary systems can be placed on moorings or autonomous vehicles and, with confidence, be left in the water for months. The result is that our ocean hydrographic database is vastly expanded in terms of both vertical resolution and of the number of places and times sampled.

Similarly, moored current measurements were beginning to be made routinely in the 1960s, but there were distinct limitations in terms of deployment length (a couple of months, perhaps) and where they could be deployed (not too close to the ocean surface). Modern moored instrument systems have since vastly improved in terms of accuracy, reliability, flexibility, and survivability. While some of the old constraints remain (such as biofouling and resolving conflicts with the fishing community), spatial and temporal coverage of high-quality moored measurements (velocity, temperature, salinity, along with interdisciplinary observations) has improved enormously. Further, acoustic current profilers deployed from moving platforms (e.g., ships and gliders) have also entered the mix. In addition, shore-based radar measurements of surface ocean currents provide unprecedented spatial characterizations. The flood of new information has been augmented over the last two decades by the development of systematic, needs-oriented observing systems, as well as by sustained science-oriented ocean observatories.

In the mid-1960s, remote sensing of the ocean was still in the developmental stage. I recall vividly the first time I saw an image of sea surface temperature, probably in the early 1970s. It was an amazing revelation that quickly answered some questions but left bigger ones. It became obvious immediately, if it had not been already, that the ocean was rich with eddies and not occupied by the tidy, laminar sort of flow that some might have earlier imagined. Again, capabilities have since exploded in terms of data quality and in terms of what can be measured, such as color, winds, sea surface elevation, and salinity from satellites, or currents with land-based radar.

The stream of sustained, multiyear ocean measurements has expanded from earlier domination by a few lightship observations and coastal sea level records. Continuing observing system and observatory measurements provide an ever-growing capability to resolve processes with time scales longer than that of a single instrument deployment.

All told, much of the ocean is now observable. We can reach locations, such as the upper ocean or the inner shelf, that were not accessible in the past. We can resolve new space scales, such as the shelf mesoscale[1] (around 10 km), and time scales ranging from seconds to decades. Reliable climatologies are being developed. While gaps remain (such as subsurface currents with the sort of coverage obtained by radar systems), there is a remarkable ability to observe new things about the ocean and to drive and evaluate increasingly capable numerical models. Further, our community is becoming increasingly adept at dealing with these data sets. In the mid-1960s, few oceanographers were aware of even the material in this volume's section 2.5, so some silly things could have been said and done. At present, the statistical and visualization capabilities accessible to most of us are remarkably effective.

One might begin to wonder if the data tsunami is being exploited as thoroughly as it could be. Entirely automated approaches to data quality control add capabilities, but they also might cause certain signals inappropriately to be set aside: the lesson of ocean "meddies" (e.g., McDowell and Rossby, 1978), which were ignored because they seemed

---

1 In the ocean, we usually take "mesoscale" to represent a scale comparable to the first-mode internal Rossby radius of deformation, that is, $O(NH/f)$. Thus, in the open ocean, this scale is typically several tens of kilometers or a bit more. The open-ocean submesoscale is defined by shorter scales, and often the Rossby number is $O(1)$. On the shelf, where the water is so much shallower, the mesoscale, by the Rossby radius definition, might be 5–10 km (or less during the weakly stratified winter). Even this definition is complicated by the way a sloping bottom makes this "a moving target." Numerical model results for the shelf (e.g., Brink and Seo, 2016) often suggest that the shelf *meso*scale also has $Ro = O(1)$, similar to the open-ocean *submeso*scale.

Presumably, then, the shelf mesoscale is not simply a shrunken version of the open-ocean mesoscale. All considered, in naming a scale range, one should be clear about what part of the ocean is intended.

like bad data, should always be on our minds. Further, the past practice (now distinctly quaint) of a scientist hand-drawing data presentations (sections or time series) did force people to look carefully at the detailed measurements, as opposed to our modern tendency to exploit the statistical power of large data sets without examining them in such detail. But the larger data sets are undeniably a boon to our science. Long time series, for example, allow conditional sampling (i.e., looking with statistical confidence at only a specialized subset of a data set), such as when winds and waves meet only certain conditions (e.g., Fewings et al., 2008).

The way we do observational oceanography is changing. In the 1960s, a scientist went to sea, made the measurements according to their own wisdom, conducted quality control, analyzed the measurements, and published. The scientist largely held the data, although there was an understanding that everything would be archived at some point, as indeed often happened. With our rapidly evolving technologies and observing systems, large data sets now increasingly become readily available without the need for the researcher to go to sea unless specialized, process-oriented measurements are called for. This publicly accessible approach has long been the norm in the atmospheric science community, where regular global measurements are routinely available because of the necessity for weather prediction. An individual scientist no longer "owns" most of the data. While this openness makes ocean measurements available to all, some have argued that this new mode removes the incentive for the individual to assure maximum data quality. One expects that increasingly sophisticated data processing software will fill any gap.

## 11.3.    Dynamical Concepts

Again, sophistication has grown rapidly. In the mid-1960s, most models and concepts about the coastal ocean (such as ideas about coastal upwelling) were essentially two-dimensional (offshore and depth), but that situation changed. Two particularly noteworthy landmark publications in this regard were Robinson (1964), which emphasized subinertial coastal-trapped waves, and Stommel and Leetma (1972), which made the case that observed currents over the U.S. east coast shelf could be accounted for only if an alongshore pressure gradient was present. From this point, dynamical thinking about the continental shelf became increasingly three-dimensional. Wonderful examples began to be published showing how these ideas actually worked in the real ocean (e.g., Mitchum and Clarke, 1986b).

And yet, something was missing. One is reminded of Henry Stommel's famous 1954 essay: "Why do our ideas about ocean circulation have such a peculiarly dream-like quality?". The coastal anomaly was that there was a body of theory that envisioned an ocean that varies only slowly, over hundreds or thousands of kilometers, alongshore and that produced very credible results for sea level and alongshore currents. Yet, this same theory worked terribly for density or cross-shelf currents (e.g., Chapman, 1987). Further, observations indicated that these poorly predicted variables vary on alongshore scales of a few tens of kilometers, if not shorter (e.g., Kundu and Allen, 1976; Holladay and

O'Brien, 1975; Garvine et al., 1989). In one sense the answer was obvious: there has to be some short-scale process (or processes) which does not dominate alongshore current variance but that dominates cross-shelf current variance. The problem was identifying the process(es?) and documenting it (them?) with observations.

Eventually, two classes of phenomena were isolated and explored. One was the instability of shelf currents (chapter 9, e.g., Flagg and Beardsley, 1978; Barth, 1994; Chapman and Gawarkiewicz, 1997; Durski and Allen, 2005, to name just a few). These studies were increasingly dominated by process-oriented numerical models and led to an appreciation that it would be hard to find many places over the shelf that are not at least sometimes susceptible to instabilities and thus the generation of short-scale currents. Second, flow features, such as mesoscale eddies, in the adjoining deep ocean were shown to affect currents at the outer edge of the shelf and so to create highly localized cross-shelf currents (chapter 10; e.g., Davis, 1985; Beardsley et al., 1985; Strub et al., 1991; Cherian and Brink, 2018). The growing understanding of these phenomena naturally involved a greater appreciation of the underlying nonlinearity of the processes, particularly of the coupling of so many phenomena over the shelf. For example, large-amplitude internal waves generate turbulent mixing (e.g., Lamb, 2014), which, in turn, can affect a range of phenomena, including coastal-trapped waves and upwelling. Thus, our perception of the coastal ocean is less dreamlike than it had been. If there was ever a time when continental shelf theoreticians thought that their ocean obeys only linear dynamics, the limitations to that belief are now being appreciated.

Although this summary emphasizes midshelf wind-driven currents and their consequences, parallel tales of increasing dynamical sophistication could be told about other aspects of flow over the shelf. Topics such as buoyancy currents (e.g., Nof and Pichevin, 2001; Horner-Devine et al., 2015) and wave-driven processes on the inner shelf/surf zone (e.g., Kumar and Feddersen, 2017) have followed similar enriching trajectories in which the interplay of observations and theory was critical.

## 11.4.  Numerical Models

In the 1960s, numerical models of shelf processes played a negligible role in our thinking, and it was only during the following decade that the models, primitive as they were by present standards, started to be influential for coastal oceanography. Since that time, a host of developments have made numerical models more and more useful. The obvious change, of course, was the increasing capability (e.g., speed and memory) of the computers themselves, but there was far more involved. Improved mixing parameterizations, data assimilation, better treatment (or avoidance) of open boundaries, model nesting, and many other developments led to better ocean simulations (e.g., Pinardi et al., 2017) and an increasing acceptance by the broader community.

Expanding capabilities broaden the range of problems where the models could be applied. For example, improved mixing parameterizations allow the turbulent inner shelf to be treated with confidence, and model nesting enables studies of how basin-scale processes affect flow over the shelf. Displays of model outputs now have the spatial and temporal richness that we are accustomed to seeing, for example, in satellite temperature images. Thus, we all need to fight the temptation to treat a model as realistic simply because it has the complex spatial structure that we have come to expect based on remote sensing.

In the context of this volume on ocean dynamics, the most influential models have been those that treat an idealized, controlled problem to shed light on some process. Models of this sort allow the user to go far beyond the limitations of analytical theory and so ask questions about the evolution of shelf eddy fields, for example. The increasingly realistic parameter range that these studies represent allow comparisons with observed conditions and so a greater degree of applicability to the real world. Extension to more realistic simulations can further allow practical predictions and a rigorous evaluation of model fidelity.

Much is yet to be learned. For example, how fine does model resolution need to be to resolve the shelf mesoscale (i.e., scales of less than about 10 km)? This may sound straightforward, but synoptic measurements on this scale are exceedingly difficult, especially if there is a need to resolve larger-scale shelf-wide processes and the shelf mesoscale simultaneously. If the observations do not yield an adequate characterization, then it is not clear, at least from the data, what range of scales a model must resolve for some particular problem that involves scale interactions. Say, for example, that one is interested in a model of wind-driven upwelling (scale of tens of kilometers) in the presence of a river-driven buoyancy current (scales of a few kilometers) while resolving the shelf mesoscale (scales of perhaps 10 km down to an undetermined level). Careful evaluations of model studies ought to resolve many modeling issues of this sort.

## 11.5.    Processes

Over the decades, the problems of greatest interest have shifted dramatically. During the 1960s and 1970s, the topics receiving the most attention included coastal upwelling, mean flow generation over the shelf, tides, and coastal-trapped waves. Attention was focused on the midshelf (water depths of perhaps 30–200 m) and generally below the hard-to-observe upper 20 m or so. Dynamical thinking was predominantly linear (or nearly so) and was dominated by analytical theories. With time, the database grew, and much of what could be done with the linear theory of these processes was exhausted. Newer problems, which were increasingly demanding observationally and difficult mathematically, came to the forefront. Attention drifted away from the midshelf and gravitated toward newly accessible processes involving buoyancy currents, the inner shelf, eddies, and internal tides, for example.

What about cross-shelf exchange? In section 1.4, this was named as a central theme. Indeed, almost every chapter (chapter 5 is the exception) in this volume at least touches on this topic. Clearly, no single process dominates this exchange: mechanisms vary from place to place and time to time. Some of the processes involved have been studied intensively, others less so. In most regards, there is more to learn, such as understanding the source depth from which upwelling water arises or predicting the eddy fluxes associated with the many forms of instability possible over the shelf. Much progress has been made over the last decades, but there is still much to be done.

The overall need to understand processes in the coastal ocean will not vanish. The motivations listed in chapter 1 are not going away and most are likely to become more pressing over the coming decades. Dealing with these needs will require more observations, improved numerical models, and continued advances in understanding underlying processes. Failure to understand these fundamentals can lead (and has led, in my experience) to failures in using numerical models or interpreting observations. Proper application of the fundamentals will deepen understanding while facilitating modeling, prediction, and effective observations.

The continuing evolution in scientific focus is exactly what one expects in a lively, active research field. A volume such as this one will surely be seen, within a decade or less, as missing important recent findings. Indeed, it is certain that new observations and new model studies will lead to radical new insights and directions. However, the expectation is that the core materials treated here will remain relevant even as frontiers continue to advance, and they assuredly will do so.

# Appendix

# Exercises

## A.1.  Introduction

The following exercises are meant to provide some insights into a few coastal ocean processes, as well as to give some experience with typical derivations. This small collection does not come close to covering the entire range of topics in this volume. With each exercise, there is also a "solution" which provides an outline to solving the problem.

## A.2.  Slab-like Mixed Layer (Chapter 3)

Consider a surface Ekman layer, where the eddy viscosity is

$$A = A_0 \qquad \text{for } z > -h_{ML,} \tag{A.2.1a}$$

$$A = 0 \qquad \text{for } z < -h_{ML}. \tag{A.2.1b}$$

Consider steady flow and say that

$$-f v_E = A u_{Ezz,} \tag{A.2.2a}$$

$$+f u_E = A v_{Ezz}. \tag{A.2.2b}$$

At the ocean surface,

$$\rho_0 A u_{Ez} = \tau_0^x, \tag{A.2.3a}$$

$$A v_{Ez} = 0. \tag{A.2.3b}$$

Solve for $u_E$ and $v_E$. Then, take the limit as $A_0$ becomes large (in what sense?) and describe what happens.

*Hint*: Stress must be continuous at the bottom of the layer, and it is zero below $z = -h_{ML}$ (since $A = 0$ there). Thus, stress $= 0$ at the bottom of the layer ($z = -h_{ML}$).

***Solution:*** The final result (after expanding for small $h_{ML}/\delta$, where $\delta = \sqrt{2 A_0 / |f|}$) is that

$$(u_E, v_E) = (0, -\tau^x)/(\rho_0 f h_{ML}). \tag{A.2.4}$$

## A.3.  Bottom Stress Formulations (Chapter 3)

First, solve the bottom Ekman layer problem for a constant eddy viscosity:

$$-fv_E = A_0 u_{Ezz,} \tag{A.3.1a}$$

$$+fu_E = A_0 v_{Ezz}. \tag{A.3.1b}$$

The interior velocity is $(u_I, v_I)$ and $u_E \to 0$ and $v_E \to 0$ as $z \to \infty$. Also, $u_I + u_E = 0$ and $v_I + v_E = 0$, at the bottom, $z = 0$. Derive an expression for bottom stress as a function of interior velocity. Also derive an expression for Ekman transport as a function of bottom stress.

Then, derive an expression for the vertical velocity at the top of the bottom boundary layer in terms of the bottom stress. You can do this directly from (A.3.1) and the continuity equation.

Sometimes, people choose to make life simple and express bottom stress as $\tau_B^x = \rho_0 \sigma_F u_I$ and $\tau_B^y = \rho_0 \sigma_F v_I$. Find an expression for $\sigma_F$ in terms of $f$ and $A_0$. What do you think has been lost by this simplification? Do you think it is important?

**Solution:** The bottom stress in terms of interior velocity can be found in section 3.3. Integrating (A.3.1) vertically, and using the definition of stress leads to

$$V_E = \tau_B^x/(\rho_0 f), U_E = -\tau_B^y/(\rho_0 f). \tag{A.3.2}$$

If you integrate the continuity equation vertically over the bottom boundary layer, you get

$$w_{z=\infty} = -U_{Ex} - V_{Ey}. \tag{A.3.3}$$

If (accounting consistently for veering for $f > 0$) $\tau_B{}^x = au_I - bv_I$ and $\tau_B{}^y = av_I + bu_I$,

$$w|_{z=\infty} = [b(u_{Ix} + v_{Iy}) + a(v_{Ix} - u_{Iy})]/(\rho_0 f). \tag{A.3.4}$$

For the problem as done up to here, $a = b = \rho_0 \sqrt{Af/2}$, and for the new form, $a = \rho_0 \sigma_F$, $b = 0$. The $b$ term (related to stress perpendicular to the interior flow, hence transport parallel to the interior flow) is proportional to the horizontal divergence of the interior flow (A.3.4). This is generally small for rotating flows that are slowly changing in time (it is exactly zero if the flow is geostrophic on an $f$ plane). The $a$ term is proportional to relative vorticity and so can be expected to be important. Thus, you would expect your answers not to depend much on $b$, so you could often get away with setting it to zero. By and large, the $b$ term should be small for frequencies lower than the inertial.

## A.4.  Damped Shelf Waves (Chapter 5)

Consider a coast at $x = -L$, where the depth $h$ varies offshore ($H_1 < H_0$) as

$$h = H_1 \text{ for } x < 0, \qquad \text{and} \tag{A.4.1a}$$

$$h = H_0 \text{ for } x > 0. \tag{A.4.1b}$$

The depth-integrated equations of motion are

$$U_t - fV = -gh\zeta_x - \sigma_F \, U/h, \tag{A.4.2a}$$

$$V_t + fU = -gh\zeta_y - \sigma_F \, V/h, \tag{A.4.2b}$$

$$\zeta_t + U_x + V_y = 0, \tag{A.4.2c}$$

where $\sigma_F$ is a constant frictional coefficient.

a. Neglect bottom friction, and then present an argument for making the rigid-lid approximation. Base your argument on terms in the vorticity equation. Now make the approximation.
b. Include friction, and derive a shelf wave dispersion relation $\omega = \omega(l, \sigma_F, \ldots)$ where

$$\zeta = \zeta_0(x) \exp[i(\omega t + ly)]. \tag{A.4.3}$$

   *Hint*: You will need to match pressure and cross-shelf transport at the topographic jump.
c. Define the "coastal long-wave" approximation and apply it to the dispersion relation. You can also assume that $H_1 \ll H_0$.
d. What do you conclude about the relative importance of friction over the shelf compared with in the deep ocean?
e. Under what conditions could you neglect friction in this problem?

*Solution:* Given depth profile (A.4.1), with a coast at $x = -L$, (A.4.2) with $\sigma_F = 0$, and (A.4.3), write a single equation for $\zeta_0(x)$:

$$0 = \omega(h\zeta_{0x})_x + (-\omega h l^2 + h_x f l)\,\zeta_0 - \omega\,(f^2 - \omega^2)\zeta_0/g. \tag{A.4.4}$$

a. If you scale this, comparing the last term with the first, you get that the last term is

$$O[f^2 L^2/(gh)] \tag{A.4.5}$$

relative to the first. This ratio compares a typical scale $L$ (shelf width) with the barotropic radius of deformation $(\sqrt{gh})/f$. The latter scale might typically be $O(2000 \text{ km})$, and the former $O(100 \text{ km})$ or less. So, the last term (due to free surface motion) is tiny in most cases.

b. To solve this, note that $h_x = 0$ over both the shelf and slope and use (A.4.3), so the problem reduces to

$$\zeta_{0xx} - l^2\,\zeta_0 = 0 \tag{A.4.6}$$

in both the shallow and deep regions. Thus,

$$\zeta_0(x) = B\exp(-lx) + C\exp(lx) \qquad \text{for } x < 0, \tag{A.4.7a}$$

$$\zeta_0(x) = b \exp(-lx) \qquad\qquad \text{for } x > 0. \tag{A.4.7b}$$

This reduces to matching things up at $x = 0$. Use

$$U = 0 \qquad\qquad \text{at } x = -L, \tag{A.4.8a}$$

$$\zeta_0 \text{ continuous} \qquad\qquad \text{at } x = 0, \tag{A.4.8b}$$

$$U \text{ continuous} \qquad\qquad \text{at } x = 0. \tag{A.4.8c}$$

This last condition, expressed in terms of $\zeta_0(x)$ is

$$(f^2 - \omega_0^2)[\omega_1 \zeta_{0x} + l f \zeta_0] H_1 = (f^2 - \omega_1^2)[\omega_0 \zeta_{0x} + l f \zeta_0] H_0, \tag{A.4.9a}$$

where

$$\omega_1 = \omega - i\sigma_F / H_1, \qquad\qquad \omega_0 = \omega - i\sigma_F / H_0. \tag{A.4.9b,c}$$

Then

$$[f(H_0 - H_1) - H_1 \omega_0] \tanh(lL) = H_0 \omega_1. \tag{A.4.10}$$

c. The coastal long-wave approximation requires that $lL \ll 1$; therefore, $\tanh(lL) \approx lL$. You can then assume that $\omega_0 / f \ll 1$, $\sigma_F / (fH_0) \ll 1$, and $H_1 \ll H_0$, so that

$$\omega = f(H_0 - H_1) \, lL \, / \, H_0 + i\sigma_F / H_1. \tag{A.4.11}$$

Notice that the inviscid, coastal long-wave phase speed reduces to (5.7.21):

$$c = f(H_0 - H_1) \, L \, / \, H_0. \tag{A.4.12}$$

This is a handy thing to remember: long shelf wave phase speed is proportional to $fL$, and the fractional depth change.

d. Friction appears in (A.4.11) only as $i\sigma_F / H_1$. The imaginary part of frequency does not depend on the deep-ocean water depth. From this we can conclude that damping counts only in the shallower water, which makes sense, since the Ekman number is much greater there (the boundary layer occupies a greater fraction of the water column there).

e. When can you neglect friction? When $\sigma_F / (\omega H_1) \ll 1$. Note that the important frequency here is $\omega$, not $f$. Thus, an Ekman number is not the correct factor to consider for this purpose.

## A.5.    Forced Shelf Wave Derivation (Chapter 5)

Consider a straight coast (along $x = 0$), with arbitrary water depth $h(x)$. The depth $h$ reaches a constant value far offshore. Upward and alongshore are $z$ and $y$, respectively. Ignore density stratification, bottom friction, and divergence associated with free-surface displacement. Making the coastal long-wave approximation, the linearized equations of motion (depth averaged) are

$$fv = g\zeta_x, \tag{A.5.1a}$$

$$v_t + fu = -g\zeta_y + \tau_0^y/(\rho h), \tag{A.5.1b}$$

$$(uh)_x + (vh)_y = 0. \tag{A.5.1c}$$

Assume that the alongshore wind stress $\tau_0^y$ varies in time and the alongshore direction but not in $x$.

Given this as a starting point, derive a forced, first-order wave equation that governs the response of the wave modes, as defined by *stream function* (chapter 5 does this in terms of pressure).

To do this:

a. Derive a single equation for stream function.
b. Find the equations that define the free modes in terms of stream function. This means developing a suitable offshore boundary condition.
c. Derive the orthogonality condition.
d. Go back to the forced problem and expand in terms of the free stream function modes.

**Solution:** This problem is essentially the derivation in Gill and Schumann (1974).

Define

$$vh = \psi_x \text{ and } uh = -\psi_y. \tag{A.5.2a, b}$$

Then, you can get a single equation for $\psi$:

$$[\psi_{xt}/h]_x - f\psi_y(1/h)_x = (\tau_0^y/\rho)(1/h)_x. \tag{A.5.3}$$

Now look for free modes like

$$\psi = \varphi(x)\, F(y,t), \tag{A.5.4}$$

with

$$\psi = 0 \quad \text{so } \varphi = 0 \quad \text{at } x = 0 \quad \text{and} \tag{A.5.5a}$$

$$\psi_x \to 0 \quad \text{so } \varphi_x \to 0 \quad \text{as } x \to \infty. \tag{A.5.5b}$$

The second condition comes from assuming that the bottom is flat far from shore ($h_x = 0$), so $\psi_{xx} = \varphi_{xx} = 0$ far offshore. Then, since $\varphi$ must be bounded, $\varphi$ must just be a constant far from shore.

Using (A.5.4) in equation (A.5.3) yields

$$[\varphi_{xt}/h]_x - f\varphi c^{-1}(1/h)_x = 0, \tag{A.5.6a}$$

$$cF_y - F_t = 0, \tag{A.5.6b}$$

where $c$ is the separation constant (phase speed of free waves). There is an infinite set of these free wave solutions, so they are labeled with a subscript $n$.

The next step is to derive an orthogonality relation. Multiply equation (A.5.6a) for the $n$ mode by $\phi_m$, and multiply equation (A.5.6a) for the $m$ mode by $\phi_n$. Subtract these and integrate the result from $x=0$ to $\infty$, then use the boundary conditions to simplify. The result is

$$0 = (1/c_n - 1/c_m) f \int (h_x/h^2) \, \phi_n \, \phi_m \, dx. \tag{A.5.7}$$

Say that

$$\delta_{nm} = f \int (h_x/h^2) \, \phi_n \, \phi_m \, dx. \tag{A.5.8}$$

Now, approach the forced problem by saying that

$$\psi = \Sigma \, \phi_n(x) \, Y_n(y,t). \tag{A.5.9}$$

Substitute this expression into (A.5.3), multiply by $\phi_m$, and integrate with regard to $x$. Then obtain (after some manipulation)

$$\Sigma \left[ -Y_{nt} c_n^{-1} f \int (h_x / h^2) \, \varphi_n \varphi_m \, dx + Y_{ny} f \int (h_x / h^2) \varphi_n \varphi_m \, dx \right] = (\tau_0^y / \rho) \int \varphi_m (1/h)_x \, dx \bigg] \tag{A.5.10}$$

Using (A.5.8), all the terms with $n \neq m$ vanish, and this becomes

$$c_m^{-1} Y_{mt} - Y_{my} = b_m (\tau_0^y / \rho), \tag{A.5.11}$$

where the wind coupling coefficient is

$$b_m = \int \phi_m \, (h_x/h^2) \, dx. \tag{A.5.12}$$

Gill and Schumann came to the same conclusion, but how they included the forcing can seem a bit mysterious.

## A.6.     Tides in a River (Chapter 7)

Consider a long, narrow, straight river along the $x$ axis with uniform depth $H$, width $W$, and length $L$. The linearized equations of motion are then

$$U_t = -gH \, \zeta_x - \sigma_F U/H \tag{A.6.1a}$$

and

$$\zeta_t + U_x = 0, \tag{A.6.1b}$$

where $\zeta$ is the free-surface elevation, $U$ is the depth-integrated upstream velocity, and $\sigma_F$ is a frictional coefficient.

Say the river mouth is at $x=0$ and there is a tidal forcing of the form

$$\zeta = B \exp(i\omega t) \qquad \text{at } x=0, \tag{A.6.2}$$

and say there is a dam far upstream, so that $U=0$ at $x=L$.

1. Why can we ignore rotation here? (Give a scaling argument.)
2. Say that $\zeta = Z(x)\exp(i\omega t)$ and solve for the complex amplitude $Z$.
3. Under what circumstances does the river's response look like a standing wave, and when does it look like a propagating wave? Express the answer as a nondimensional number. Explain in physical terms.
4. One might question whether the linear assumption is a good idea for realistic circumstances. Where do you think the linearity assumption might break down (there are several possibilities), and what would be the physical implications? Assume, for a reasonable river, that $H$ is in the range of 1–10 m.

***Solution:***

1. Why can we ignore rotation here?

Consider the upstream momentum equation

$$U_t - fV = \ldots, \tag{A.6.3}$$

where $V$ is the cross-stream transport. The question is: When can the $fV$ term be dropped? From continuity,

$$U_x = O(V_y), \tag{A.6.4}$$

so that

$$V = O[(W/L^x)U], \tag{A.6.5}$$

and so you can neglect $fV$ relative to $U_t$ if

$$\omega \gg f(W/L^x), \tag{A.6.6}$$

where $\omega$ is the same order as $f$ at tidal frequencies. This does not mean that rotation is unimportant in the other (cross-stream) momentum equation.

2. The solution is

$$Z(x) = B\cos[l(x+L)]/\cos(lL), \tag{A.6.7}$$

where

$$l^2 = \omega(\omega - i\sigma_F/H)/(gH). \tag{A.6.8}$$

3. If $\sigma_F = 0$, then $l$ is real, and $Z$ is in phase (or precisely out of phase) at all $x$ locations ($Z(x)$ is real). If $l$ is complex, then there is an along-channel phase gradient (net propagation).

Physically, the ingoing wave is more energetic that the damped, outgoing wave, so the inward propagation dominates over the damped, reflected propagation.

With this in mind, you could say that the time scale for wave propagation along the channel ($T_F = L/\sqrt{gH}$) has to be short relative to the dissipation time scale ($T_D = H/\sigma_F$) to ignore along-channel phase differences; that is,

$$\sigma_F L / (H\sqrt{gH}) \ll 1. \tag{A.6.9}$$

Alternatively, you could simply ask when friction in the momentum equation can be neglected, so that, equivalently,

$$\sigma_F/(\omega H) \ll 1. \tag{A.6.10}$$

4. There are several potential nonlinearities:
   - Bottom stress: You may need to use the quadratic form if you want a detailed solution.
   - Free-surface elevation: If $\zeta$ is not small compared with $H$, you may get several nonlinear wave phenomena: bores, breaking waves, and so forth.
   - Momentum advection can come into play if $U/(\omega HL)$ is not small.

In any of these cases, the response is no longer a single sinusoid but involves coupling with other frequencies.

## A.7.    Data Analysis: Tides and Winds

The data files on this book's website contain hourly data (sf_23m_hr_hw.dat) and low-pass filtered data (tides removed: sf_23m_hw.dat). The current meter data are for the period 1998–1999 at the Georges Bank "Southern Flank" location on map A.1. Wind stress (txhw, tyhw) is from the same location and is in dyne/cm². Sea level (zhw) is from Boston and is in cm. (Boston is just off the map to the west.) Currents (uhw, vhw, uuhw, vvhw) are in cm/s. All the vectors are in (east, north) coordinates. Time (thr, thw) is in 1995 days.

1. Make a sensible coordinate rotation for the vector time series. Explain your choice, preferably in terms of the statistics of the flow field.
2. Using least-squares fits, find the amplitudes of the $M_2$ and $K_1$ tides for both $u$ and $v$.
3. Using correlations (or some other statistical tool) and the low-pass-filtered data, explore the relations between currents, winds, and sea level. Explain what you find in physical terms.

***Solution:***
1. The new coordinate system, having velocity components $(u', v')$, is found by rotating the original (east, north, i.e., $u$, $v$) velocity counterclockwise through an angle $\theta$. The rotated velocity is given by

$$u' = u \cos \theta + v \sin \theta, \tag{A.7.1a}$$

$$v' = -u \sin \theta + v \cos \theta, \tag{A.7.1b}$$

and $\theta$ (roughly 20°) can be chosen so that $u$ is along isobaths. The standard deviations of the low-pass-filtered velocities will be much larger in the along-isobath direction than cross-isobath.

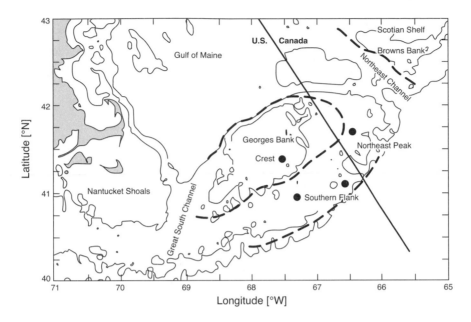

**FIGURE A.1.** A locator map for the measurement locations in this exercise. Adapted from Brink et al. (2009).

2. Do least-squares fits on the hourly data in the form

$$u_F = a_1 \cos \omega_1 t + a_2 \sin \omega_1 t + a_3 \cos \omega_2 t + a_4 \sin \omega_2 t + a_5, \tag{A.7.2}$$

solving for the $a_n$. This is done by minimizing the error:

$$Error = \Sigma \, (u - u_F)^2 \tag{A.7.3}$$

(where the sum is over time) with regard to the coefficients. Thus, find a set of five equations based on $\partial Error/\partial a_1 = 0$, and so on.

3. Using the low-pass-filtered data, look for correlations between sea level and along-bank currents, and between wind components and sea level and currents. Allow for the possibility that some variables may be out of phase with each other (i.e., leading or lagging in time). What do these findings tell us?

# Glossary

**Adiabatic:** Concerning processes that conserve energy.

**Aliased:** Describing data that are sampled too infrequently in time and/or space. Useful information is missed between the measurements.

**Anisotropic:** Describing quantities whose statistical properties (e.g., standard deviation) differ depending on spatial direction.

**Anoxia:** The state of having depleted dissolved oxygen concentrations in the ocean.

**Baroclinic:** Having density and pressure surfaces that are not parallel. Most commonly taken to mean that there are depth variations in currents that occur associated with density gradients, as opposed to with turbulent mixing processes.

**Baroclinic instability:** The growth of disturbances due to the extraction of energy from the initial gravitational potential energy associated with horizontal density differences.

**Baroclinic modes:** An infinite set of mathematically orthogonal solutions to a vertical structure equation in a density-stratified ocean. Their existence requires the surface and bottom to be flat, along with other conditions. The modes are found as eigenfunctions of a differential equation problem. See section 5.4.

**Barotropic:** Having density and pressure surfaces that are parallel. Most commonly taken to mean that the horizontal velocity components do not vary with depth.

**Barotropic instability:** The growth of disturbances due to extraction of energy from the initial kinetic energy field. Specifically, this instability feeds off horizontal velocity shears, but it does *not* require that the flow be barotropic.

**Basin mode:** A resonant standing wave, having a defined frequency, in an oceanic region.

**Bathymetry:** The topography of water depth.

**$\beta$:** The gradient of planetary vorticity $f$ due to the curvature of Earth's surface.

**$\beta$ plane:** An approximation that retains Cartesian coordinates and treats the Coriolis parameter $f$ as a linear function of the north-south coordinate.

**Bore:** A highly nonlinear wave that is sufficiently abrupt to be effectively a traveling discontinuity. One example is a tidal bore: a sharp transition from shallow to deeper water that propagates upstream in a river.

**Boundary layer:** In the coastal context, a portion of the water column adjoining either the surface or the bottom. Very often, the regions are highly turbulent (relative to waters outside the boundary layer) and thus relatively well mixed.

**Coastal-trapped wave:** A propagating disturbance which is constrained to translate in the alongshore direction and which decays with distance offshore. The trapping generally occurs because of topographic or rotational constraints (or a combination of the two). Internal Kelvin waves and continental shelf waves are two subinertial frequency limits of this more general class, corresponding to relatively strong and weak density stratification, respectively.

**Continental shelf wave:** A barotropic wave that expresses the fluctuating exchange between relative vorticity (particle spin) and vorticity associated with water depth changes on a rotating planet. Stated another way, the underlying physics involves water parcel stretching or contracting, with vorticity conservation requiring that its spin change.

**Dirac delta function:** A mathematical function $\delta_D(\xi)$ that represents an infinite, narrow spike. It is equal to zero everywhere except where $\xi = 0$. Further, its integral over all $\xi$ is exactly 1.

**Dispersion relation:** An equation that relates the space and time scales (i.e., the frequency and wavenumbers) of a monochromatic wave.

**Dispersive:** Describing waves whose phase or group velocities change when the wavelength changes. Given enough time, a linear dispersive wave packet will break up into a range of sinusoids with varying frequencies.

**Ecliptic:** Referring to the plane defined by Earth's orbit around the sun.

**Eddy viscosity:** A quantity $A$ that is meant to characterize the strength of turbulent dissipation, so that a stress (averaged over many turbulent fluctuations) can be written in the form $\tau = \rho_0 A v_z$, where $\rho_0$ is the fluid density, and $v_z$ is a current shear. Ultimately, the eddy viscosity is an empirical construct, but it is possible to make very credible estimates using turbulence closure models. Eddy diffusivity is an analogous quantity that relates to turbulent diffusion rather than stress.

**Edge wave:** A class of coastal-trapped wave that exists over sloping topography and does not require the existence of rotation or density stratification. Trapping occurs because of refraction associated with the local increase in speed of the gravity wave in deeper water. Its frequency is always greater than the inertial.

**Eigenvalue/eigenfunction:** In a number of linear mathematical problems (both differential equations and matrix problems), a set of functions (eigenfunctions), each paired with a number (eigenvalue) that is, in a certain sense, associated with the function. The baroclinic modes of section 5.4 (equations 5.4.10) are an example. When this set of functions exists, it has important mathematical properties that make it useful for solving more complicated problems (e.g., section 5.7).

**Equatorward:** The alongshore direction tending away from the North Pole in the northern hemisphere, or South Pole in the southern hemisphere.

**Eulerian velocity:** Fluid motion as observed at a fixed location, in contrast with Lagrangian velocity.

*f* **plane:** An approximation involving a Cartesian coordinate system that treats the locally vertical component of Earth's rotation as a constant.

**Friction velocity:** The quantity $u_* = (\tau/\rho)^{1/2}$, where $\tau$ is a turbulent stress, and $\rho$ is a fluid (air or water, for example) density. This can be thought of as a representative speed of a turbulent fluctuation.

**Front:** A sharp horizontal change in some quantity (such as surface temperature) that can sometimes even be thought of as a discontinuity.

**Geostrophic:** Steady, linear, inviscid flow in a rotating system is generally in geostrophic balance; that is, the horizontal pressure gradient is balanced by the Coriolis force.

**Geostrophic adjustment:** The process by which the mass and momentum fields come into geostrophic balance. For example, the initial state might be a given velocity field in which pressure is constant everywhere. Both pressure and the velocity field would then change until they achieve a state in geostrophic balance.

**Group velocity:** The vector describing how quickly and in what direction a linear wave carries energy.

**Hypoxia:** A state of ocean oxygen concentration so low that it adversely affects living conditions for animals.

**Independence time (or length) scale:** The lag time (distance) interval over which a time (spatial) series is correlated with itself. It is usually defined as the integral of the auto-correlation function.

**Inertial oscillation:** A circularly oscillating motion at the frequency $f$.

**Inverse barometer:** The assumption that when air pressure changes over a portion of the ocean, sea surface height adjusts so that there is no gradient of subsurface pressure.

**Isobath:** A contour of constant water depth.

**Isotropic:** Referring to properties that vary similarly in all spatial directions, as opposed to anisotropic.

**Kelvin wave:** A coastal-trapped wave, found in its purest form in a rotating ocean with a flat bottom and a vertical coastal wall. It propagates nondispersively at a long gravity wave speed and decays exponentially with distance offshore.

**Lagrangian velocity:** Fluid motion measured following a water parcel wherever it goes, in contrast with the Eulerian velocity.

**Meridional:** In the north-south direction.

**NPZ:** Shorthand for "nutrient-phytoplankton-zooplankton," describing simple biological models that include only these three compartments.

**Persistence:** The tendency for something to remain nearly the same. Here, it is used in the context of the independence time scale. Larger time scales (slower variations) mean more persistence.

**Phase velocity:** The rate at which wave crests or troughs move. Only when the wave is nondispersive is this the rate at which energy is transmitted.

**Phytoplankton:** Microscopic plants living in the ocean, for example.

**Planetary vorticity:** A particle's spin associated with the rotation of the platform (such as Earth) on which it is located. Equal to the Coriolis parameter $f$.

**Plankton:** Very small, often microscopic living things (for example, plants: phytoplankton; animals: zooplankton; fish: ichthyoplankton; bacteria: bacterioplankton) whose motion is largely passive, determined by the ambient currents.

**Poincaré wave:** A long gravity wave that is modified by the planet's rotation.

**Poleward:** The alongshore direction that would bring the observer closer to the North Pole in the northern hemisphere (or to the South Pole in the southern hemisphere).

**Rectification:** The process by which a fluctuating (zero mean) flow can generate a steady (time mean) current.

**Rigid lid:** An approximation, valid for smaller-scale (relative to the barotropic Rossby radius) flows, that neglects the mass storage associated with vertical motions of the free surface. Free-surface motions remain important for establishing pressure gradients.

**Rip current:** A strong offshore flow in the surf zone that is concentrated into a narrow jet.

**Rossby radius:** The natural length scale that occurs for divergent flows on a rotating planet. It is expressed as a gravity wave speed (either barotropic or internal) divided by the Coriolis parameter (a measure of the planet's effective rotation rate at a given latitude).

**Rossby wave:** A large-scale wave in the ocean or atmosphere that expresses a fluctuating exchange between planetary vorticity (associated with Earth's rotation and varying with latitude) and the relative vorticity of a water parcel.

**Shear dispersion:** The process by which a vertically sheared horizontal flow combines with vertical mixing to cause horizontal spreading.

**Shelf wave:** See Continental shelf wave.

**Sidereal day:** The length of a day (about 23 hours and 56 minutes) defined relative to distant, "fixed" stars. The familiar 24-hour solar day is defined by the time from noon until noon and is slightly longer because it accounts for Earth's motion around the sun.

**Stokes drift:** The net motion of a water parcel moving in a wave field whose amplitude varies systematically with location. For example, short (nonhydrostatic) gravity waves have larger current fluctuations near the surface, so a water parcel experiences a stronger motion in the direction of propagation when it is displaced upward than the weaker opposite motion when it is displaced downward.

**Surf zone:** That shallow part of the coastal ocean, a few meters deep, characterized by breaking gravity waves.

**Synoptic:** Describing measurements that are made quickly enough relative to the natural variability that they can be taken to be effectively simultaneous (i.e., a "snapshot").

**Syzygy:** The state in which Earth, the Moon, and the Sun are all located along a single straight line. This alignment is conducive to particularly strong tidal forcing. A really great word, from the Greek, meaning "yoked together."

**Thermal wind:** Describing flow that is geostrophic and vertically sheared due to horizontal density gradients.

**Vorticity:** A measure of the spin of a fluid parcel. It can be associated with shearing motion, curving trajectories, or with the rotation of the coordinate frame.

**Wavenumber:** A measure of a wave's length scale. Specifically, it is $(2\pi)$ divided by the wavelength in the direction of a particular axis; for example, $k = 2\pi/$ (the wavelength in the $x$ direction).

**Wave setup:** Piling up of water near the coast driven by the onshore flux of wave momentum.

**Zonal:** In the east-west direction.

**Zooplankton:** Small, usually microscopic, animals living, for example, in the ocean, whose location is largely determined by ambient fluid motions.

# References

Abramowitz, M., and I.A. Stegun, 1965. *Handbook of Mathematical Functions*. Dover Publications, New York, 1046pp.

Allen, J.S., 1976. Some aspects of the forced wave response of stratified coastal regions. *J. Phys. Oceanogr.*, 6, 113–119.

Allen, J.S., and D.W. Denbo, 1984. Statistical characteristics of the large-scale response of coastal sea level to atmospheric forcing. *J. Phys. Oceanogr.*, 14, 1079–1094.

Allen, J.S., and P.K. Kundu, 1978. On the momentum, vorticity and mass balance on the Oregon shelf. *J. Phys. Oceanogr.*, 8, 13–27.

Allen, J.S., and P.A. Newberger, 1996. Downwelling circulation on the Oregon continental shelf. Part I: Response to idealized forcing. *J. Phys. Oceanogr.*, 26, 2011–2035.

Allen, J.S., and P. Newberger, 1998. On symmetric instabilities in oceanic bottom boundary layers. *J. Phys. Oceanogr.*, 28, 1131–1151.

Allen, J.S., P.A. Newberger, and J. Federiuk, 1995. Upwelling circulation on the Oregon continental shelf. Part I: Response to idealized forcing. *J. Phys. Oceanogr.*, 25, 1843–1866.

Allen, S.E., and X. Durrieu de Madron, 2009. A review of the role of submarine canyons in deep-ocean exchange with the shelf. *Ocean Sci.*, 5, 607–620.

Anderson, D.L.T., and P.B. Rowlands, 1976. The role of inertia-gravity and planetary waves in the response of a tropical ocean to the incidence of an equatorial Kelvin wave on a meridional boundary. *J. Mar. Res.*, 34, 295–312.

Anderson, D.M., A.D. Cembella, and G.M. Hallegraeff, 2012. Progress in understanding harmful algal blooms: Paradigm shifts and new technologies for research, monitoring and management. *Annu. Rev. Marine Sci.*, 4, 143–176.

Andres, M., 2021. Spatial and temporal variability of the Gulf Stream near Cape Hatteras. *J. Geophys. Res.*, 126, e2021JC017579. https://doi.org/10.1029/2021JC017579.

Arbic, B.K., R.H. Karsten, and C. Garrett, 2009. On tidal resonance in the global ocean and the back-effect of coastal tides upon open-ocean tides. *Atmosphere-Ocean*, 47:4, 239–266, https://doi.org/10.3137/OC311 .2009.

Ardhuin, F., W.C. O'Reilly, T.H.C. Herbers, and P.F. Jessen, 2003. Swell transformation across the continental shelf. Part I: Attenuation and directional broadening. *J. Phys. Oceanogr.*, 33, 1921–1939.

Austin, J.A., and J.A. Barth, 2002. Variation in the position of the upwelling front on the Oregon shelf. *J. Geophys. Res.*, 107, 3180, https://doi.org/10.1029/2001JC000858.

Austin, J.A., and S.J. Lentz, 2002. The inner shelf response to wind-driven upwelling and downwelling. *J. Phys. Oceanogr.*, 32, 2171–2193.

Badin, G., R.G. Williams, J.T. Holt, and L.J. Fernand, 2009. Are mesoscale eddies in shelf seas formed by baroclinic instability of tidal fronts? *J. Geophys. Res.*, 114, C10021, https://doi.org/10.1029/2009JC005340.

Baringer, M.O., and J.F. Price, 1997a. Mixing and spreading of the Mediterranean outflow. *J. Phys. Oceanogr.*, 27, 1654–1677.

Baringer, M.O., and J.F. Price, 1997b. Momentum and energy balance of the Mediterranean outflow. *J. Phys. Oceanogr.* 27, 1678–1692.

Barth, J.A., 1994. Short-wavelength instabilities on coastal jets and fronts. *J. Geophys. Res.*, 99, 16095–16115, https://doi.org/10.1029/94JC01270.

Barth, J.A., S.D. Pierce, and T.J. Cowles, 2005. Mesoscale structure and its seasonal evolution in the northern California Current system. *Deep-Sea Res. II*, 52, 5–28.

Barth, J.A., S.D. Pierce, and R.L. Smith, 2000. A separating coastal upwelling jet at Cape Blanco, Oregon, and its connection to the California Current system. *Deep-Sea Res. II*, 37, 783–810.

Basdurak, N.B., H. Burchard, and H.M. Schuttelaars, 2021. A local eddy viscosity parameterization for wind-driven estuarine exchange flow. Part I: Stratification dependence. *Prog. Oceanogr.*, 193, 102548.

Beardsley, R.C., D.C. Chapman, K.H. Brink, S.R. Ramp, and R. Schlitz, 1985. The Nantucket Shoals Flux Experiment (NSFE79). Part I: A basic description of the current and temperature variability. *J. Phys. Oceanogr.*, 15, 713–748.

Beardsley, R.C., C.E. Dorman, C.A. Friehe, L.K. Rosenfeld, and C.D. Winant, 1987. Local atmospheric forcing during the Coastal Ocean Dynamics Experiment 1. A description of the marine boundary layer and atmospheric conditions over a northern California upwelling region. *J. Geophys. Res.*, 92, 1467–1488.

Becherer, J., J.N. Moum, J. Calantoni, J.A. Colosi, J.A. Barth, J.A. Lerczak, J.M. McSweeney, J.A. MacKinnon, and A.F. Waterhouse, 2021. Saturation of the internal tide over the inner continental shelf. Part II: Parameterization. *J. Phys. Oceanogr.*, 51, 2565–2582.

Belcher, S.E., A.L.M. Grant, K.E. Hanley, B. Fox-Kemper, L. Van Roekel, P.P. Sullivan, W.G. Large, A. Brown, A. Hines, D. Calvert, A. Rutgersson, H. Pettersson, J.-R. Bidlot, P.A.E.M. Janssen, and J.A. Polton, 2012. A global perspective on Langmuir turbulence in the ocean surface boundary layer. *Geophys. Res. Lett.*, 39, L18605, https://doi.org/10.1029/2012GL052932.

Bell, T. H., 1975. Lee waves in stratified flows with simple harmonic time dependence. *J. Fluid Mech.*, 67, 705–722.

Blumsack, S.L., and P. Gierasch, 1972. Mars: The effects of topography on baroclinic instability. *J. Atmos. Sci.*, 29, 1081–1089, https://doi.org10.1175/1520–0469(1972)029,1081:MTEOTO.2.0.CO;2.

Book, J.W., P.J. Martin, I. Janeković, M. Kuzmić, and M. Wimbush, 2009. Vertical structure of bottom Ekman tidal flows: Observations, theory, and modeling from the northern Adriatic. *J. Geophys. Res.*, 114, C01S06, https://doi.org/10.1029/2008JC004736.

Bowen, A.J., and R.A. Holman, 1989. Shear instabilities of the mean longshore current 1. Theory. *J. Geophys Res.*, 94, 18,023–18,030.

Bowen, A.J., D.L. Inman, and V.P. Simmons, 1968. Wave 'set-down' and set-up. *J. Geophys. Res.*, 73, 2569–2577.

Bower, A.S., 1989. Potential vorticity balances and horizontal divergence along particle trajectories in Gulf Stream meanders east of Cape Hatteras. *J. Phys. Oceanogr.*, 19, 1669–1681. https://doi.org/10.1175/1520 -0485(1989)019%3C1669:PVBAHD%3E2.0.CO;2.

Bower, A.S., L. Armi, and I. Ambar, 1997. Lagrangian observations of meddy formation during a Mediterranean undercurrent seeding experiment. *J. Phys. Oceanogr.*, 27, 2545–2575.

Brasseale, E., and P. MacCready, 2021. The shelf source of estuarine inflow and its consequences for river plume shape. *J. Phys. Oceanogr.*, 51, 2407–2421.

Brink, K.H., 1980. Propagation of barotropic continental shelf waves over irregular bottom topography. *J. Phys. Oceanogr.*, 10, 765–778.

Brink, K.H., 1982. A comparison of long coastal trapped wave theory with observations off Peru. *J. Phys. Oceanogr.*, 12, 897–913.

Brink, K.H., 1986. Scattering of long coastal-trapped waves due to bottom irregularities. *Dyn. Atmos. Oceans*, 10, 149–164.

Brink, K.H., 1989. Energy conservation in coastal-trapped wave calculations. *J. Phys. Oceanogr.*, 19, 1011–1016.

Brink, K.H., 1997. Time-dependent motions and the nonlinear bottom Ekman layer. *J. Mar. Res.*, 55, 613–631.

Brink, K.H., 1999. Island-trapped waves, with application to observations off Bermuda. *Dyn. Atmos. Oceans*, 29, 93–118.

Brink, K.H., 2006. Coastal-trapped waves with finite bottom friction. *Dyn. Atmos. Oceans*, 41, 172–190.

Brink, K.H., 2010. Topographic rectification in a forced, dissipative, barotropic ocean. *J. Mar. Res.*, 68, 337–368.

Brink, K.H., 2012a. Buoyancy arrest and shelf-ocean exchange. *J. Phys. Oceanogr.*, 42, 644–658.

Brink, K.H., 2012b. Baroclinic instability of an idealized tidal mixing front. *J. Mar. Res.*, 70, 661–688.

Brink, K.H. 2013. Instability of a tidal mixing front in the presence of realistic tides and mixing. *J. Mar. Res.*, 71, 227–251.

Brink, K.H., 2016. Continental shelf baroclinic instability. Part I: Relaxation from upwelling or downwelling. *J. Phys. Oceanogr.*, 46, 551–568.

Brink, K.H., 2017. Surface cooling, winds, and eddies over the continental shelf. *J. Phys. Oceanogr.*, 47, 879–894.

Brink, K.H., 2021. Near resonances of superinertial fluctuations at islands and seamounts. *J. Phys. Oceanogr.*, 2721–2733.

Brink, K.H., R.C. Beardsley, R. Limeburner, J.D. Irish, and M. Caruso, 2009. Long-term moored array measurements of currents and hydrography over Georges Bank: 1994–1999. *Prog. Oceanogr.*, 82,191–223. https://doi.org/10.1016/j.pocean.2009.07.004.

Brink, K.H., and D.A. Cherian, 2013. Instability of an idealized tidal mixing front: Symmetric instabilities and frictional effects. *J. Mar. Res.*, 71, 425–450.

Brink, K.H., D. Halpern, A. Huyer, and R.L. Smith, 1983. The physical environment for the Peruvian upwelling system. *Prog. Oceanogr.*, 12, 285–305.

Brink, K.H., and S.J. Lentz, 2010a. Buoyancy arrest and bottom Ekman transport. Part I: Steady flow. *J. Phys. Oceanogr.*, 40, 621–635.

Brink, K.H., and S.J. Lentz, 2010b. Buoyancy arrest and bottom Ekman transport. Part II: Oscillating flow. *J. Phys. Oceanogr.*, 40, 636–655.

Brink, K.H., R. Limeburner, and R.C. Beardsley, 2003. Properties of flow and pressure over Georges Bank as observed with near-surface drifters. *J. Geophys. Res.*, 108, 8001, https://doi.org/10.1029/2001JC001019.

Brink, K.H., and H. Seo, 2016. Continental shelf baroclinic instability. Part II: Oscillating wind forcing. *J. Phys. Oceanogr.*, 46, 569–582, https://doi.org/10.1175/JPO-D-15-0048.1.

Brooks, D.A., and J.M. Bane, 1983. Gulf Stream meanders off North Carolina during winter and summer 1979. *J. Geophys. Res.*, 88, 4633–4650.

Brown, W.S., J.D. Irish, and C.D. Winant, 1987. A description of subtidal pressure field observations on the northern California continental shelf during the Coastal Ocean Dynamics Experiment. *J. Geophys. Res.*, 92, 1605–1635.

Bruner de Miranda, L., F. Pinheiro Andutta, B. Kjerfve, and B. Mendes de Castro Filho, 2017. *Fundamentals of Estuarine Physical Oceanography*. Springer, Singapore, 512pp.

Carton, J.A., 1984. Coastal circulation caused by an isolated storm. *J. Phys. Oceanogr.*, 14, 114–124.

Cartwright, D.E., 2000. *Tides: A Scientific History*. Cambridge University Press, Cambridge, U.K., 292pp.

Castelao, R.M., and J.A. Barth, 2005. Coastal ocean response to summer upwelling favorable winds in a region of alongshore bottom topography variations off Oregon. *J. Geophys. Res.*, 110, C10S04, https://doi.org/10.1029/2004JC002409.

Cazenave, A., N. Champollion, F. Paul, and J. Benveniste (eds.), 2017. *Integrative Study of the Mean Sea Level and Its Components*. Springer International, Cham, Switzerland, 416pp.

Chaigneau, A., G. Eldina, and B. Dewitte, 2009. Eddy activity in the four major upwelling systems from satellite altimetry (1992–2007). *Prog. Oceanogr.*, 83, 117–123.

Chant, R.J., S. Glenn, and J. Kohut, 2004. Flow reversals during upwelling conditions on the New Jersey inner shelf. *J. Geophys. Res.*, 109, C12S03, https://doi.org/10.1029/2003JC001941.

Chant, R.J., J. Wilken, W. Zhang, B.-J. Choi, E. Hunter, R. Castelao, S. Glenn, J. Jurisa, O. Schofield, R. Houghton, J. Kohut, T.K. Frazer, and M.A. Moline, 2008. Dispersal of the Hudson River plume in the New York Bight. *Oceanography*, 21, 148–161.

Chapman, D.C., 1984. A note on the use of two-layer models of coastally trapped waves. *Dyn. Atmos. Oceans*, 8, 73–86.

Chapman, D.C., 1987. Application of wind-driven, long, coastal-trapped wave theory along the California coast. *J. Geophys. Res.*, 92, 1798–1816.

Chapman, D.C., 2002. Deceleration of a finite-width, stratified current over a sloping bottom: frictional spin-down or buoyancy shutdown? J. *Phys. Oceanogr.*, 32, 336–352

Chapman, D.C., and G. Gawarkiewicz, 1997. Shallow convection and buoyancy equilibration in an idealized coastal polynya. *J. Phys. Oceanogr.*, 27, 555–566, https://doi.org/10.1175/ 1520–0485(1997)027,0555:SCABEI .2.0.CO;2.

Chapman, D.C., and S.J. Lentz, 1994. Trapping of a coastal density front by the bottom boundary layer. *J. Phys. Oceanogr.*, 24, 1464–1479.

Chapman, D.C., and S.J. Lentz, 1997. Adjustment of a stratified flow over a sloping bottom. *J. Phys. Oceanogr.*, 27, 340–356. https://doi.org/10.1175/1520-0485(1997)027%3C0340:AOSFOA%3E2.0.CO;2.

Chapman, D.C., S.L. Lentz, and K.H. Brink, 1988. A comparison of empirical and dynamical hindcasts of low-frequency, wind-driven motions over a continental shelf. *J. Geophys. Res.*, 93, 12,409–12,422.

Chelton, D.B., M.G. Schlax, and R.M. Samelson, 2011. Global observations of nonlinear mesoscale eddies. *Prog. Oceanogr.*, 91, 167–216.

Chen, C., R.C. Beardsley, and R. Limeburner, 1995. A numerical study of stratified tidal rectification over finite-amplitude banks. Part II: Georges Bank. *J. Phys. Oceanogr.*, 25, 2111–2128.

Chen, K., G. Gawarkiewicz, and J. Yang, 2022. Mesoscale and submesoscale shelf-ocean exchanges initialize an advective Marine Heatwave. *J. Geophys. Res.*, 127, e2021JC017927. https://doi.org/10.1029/2021JC017927.

Chen, S.-N, C.-J. Chen, and J.A. Lerczak, 2019. On baroclinic instability over continental shelves: Testing the utility of Eady-type models. *J. Phys. Oceanogr.*, 50, 3–33.

Chereskin, T.K., 1995. Direct evidence for an Ekman balance in the California Current. *J. Geophys. Res.*, 100, 18,261–18,269.

Cherian, D.A., and K.H. Brink, 2016. Offshore transport of shelf water by deep-ocean eddies. *J. Phys. Oceanogr.*, 46, 3599–3621.

Cherian, D.A., and K.H. Brink, 2018. Shelf flows forced by deep-ocean anticyclonic eddies at the shelf break. *J. Phys. Oceanogr.*, 48, 1117–1138.

Clarke, A.J., 1976. Coastal upwelling and coastally trapped long waves. PhD thesis, Cambridge University, 178pp.

Clarke, A.J., 2008. *An Introduction to the Dynamics of El Niño and the Southern Oscillation*. Elsevier, New York, 308pp.

Clarke, A.J., and D.S. Battisti, 1981. The effect of continental shelves on tides. *Deep-Sea Res.*, 28A, 665–682.

Clarke, A. J., and C. Shi, 1991. Critical frequencies at ocean boundaries. *J. Geophys. Res.*, 96, 10731–10738.

Clarke, A.J., and S. Van Gorder, 1986. A method for estimating wind-driven frictional, time-dependent, stratified shelf and slope water flow. *J. Phys. Oceanogr.*, 16, 1013–1028.

Clarke, A.J., and S. Van Gorder, 1994. On ENSO coastal currents and sea levels. *J. Phys. Oceanogr.*, 24, 661–680.

Cohen, E.B., and M.D. Grosslein, 1987. "Production on Georges Bank Compared with Other Shelf Ecosystems." In *Georges Bank*, R.H. Backus (ed.), The MIT Press, Cambridge, Mass., 383–391.

Crawford, W.R., and R.E. Thomson, 1984. Diurnal-period continental shelf waves along Vancouver Island: A comparison of observations with theoretical models. *J. Phys. Oceanogr.*, 14, 1629–1646.

Crépon, M., C. Richez, and M. Chartier, 1984. Effects of coastline geometry on upwellings. *J. Phys. Oceanogr.*, 14, 1365–1382.

Csanady, G.T., 1978. The arrested topographic wave. *J. Phys. Oceanogr.*, 8, 47–62.

Dale, A.C., J.M. Huthnance, and T.J. Sherwin, 2001. Coastal-trapped waves and tides at near-inertial frequencies. *J. Phys. Oceanogr.*, 31, 2958–2970.

Dale, A.C., D.S. Ullman, J.A. Barth, and D. Hebert, 2003. The front on the northern flank of Georges Bank in Spring: 1. Tidal and subtidal variability. *J. Geophys. Res.*, 108, 8009, https://doi.org/10.1029/2002JC001327.

Dalrymple, R.A., J.H. MacMahan, Ad J.H.M. Reniers, and V. Nelko, 2011. Rip Currents. *Annu. Rev. Fluid Mech.*, 43, 551–581.

D'Asaro, E.A., 2014. Turbulence in the upper-ocean mixed layer. *Annu. Rev. Mar. Sci.*, 6, 101–115, https://doi.org/10.1146/annurev-marine-010213-135138.

Davis, K.A., R.S. Arthur, E.C. Reid, J.S. Rogers, O.B. Fringer, T.M. DeCarlo, and A.L. Cohen, 2020. Fate of internal waves on a shallow shelf. *J. Geophys. Res.*, 125, e2019JC015377.

Davis, R.E., 1985. Drifter observations of coastal surface currents during CODE: The statistical and dynamical views. *J. Geophys. Res.*, 90, 4756–4772.

Dean, R.G., and R.A. Dalrymple, 1992. *Water Wave Mechanics for Engineers and Scientists*. World Scientific Publishing Company, New Jersey, 353pp.

deSzoeke, R.A., and J.G. Richman, 1984. On wind-driven mixed layers with strong horizontal gradients—a theory with application to coastal upwelling. *J. Phys. Oceanogr.*, 14, 364–377.

Dever, E. P., 1997. Subtidal velocity correlation scales on the northern California shelf. *J. Geophys. Res.*, 102, 8555–8572, https://doi.org/10.1029/96JC03451.

Dong, J., B. Fox-Kemper, H. Zhang, and C. Dong, 2020. The seasonality of submesoscale energy production, content, and cascade. *Geophys. Res. Lett.*, 47, 2020GL087388.

Drake, H.F., R. Ferrari, and J. Callies, 2020. Abyssal circulation driven by near-boundary mixing: Water mass transformations and interior stratification. *J. Phys. Oceanogr.*, 50, 2203–2226.

Durski, S.M., and J.S. Allen, 2005. Finite-amplitude evolution of instabilities associated with the coastal upwelling front. *J. Phys. Oceanogr.*, 35, 1606–1628.

Egbert, G.D, A.F. Bennett, and M.G.G. Foreman, 1994. TOPEX/POSEIDON tides estimated using a global inverse model. *J. Geophys. Res.*, 99, 24,821–24,852.

Egbert, G.D., and R.D. Ray, 2017. "Tidal Prediction." In *The Sea: The Science of Ocean Prediction*, Special issue, *J. Mar. Res.*, 75, 189–237.

Ekman, V.W., 1905. On the influence of the earth's rotation on ocean currents. *Ark. Mat. Astr. Fys.*, 2,1–52.

Elipot, S., and L.M. Beal, 2015. Characteristics, energetics, and origins of Agulhas Current meanders and their limited influence on ring shedding. *J. Phys. Oceanogr.*, 45, 2294–2314.

Enfield, D.B., and J.S. Allen, 1983. The generation and propagation of sea level variability along the Pacific coast of Mexico. *J. Phys. Oceanogr.*, 13, 1012–1033.

Enfield, D.B., M.D.P. Cornejo-Rodriguez, R.L. Smith, and P.A. Newberger, 1987. The equatorial source of propagating variability along the Peru coast during the 1982–1983 El Niño. *J. Geophys. Res.*, 92, 14,335–14,346.

Ezer, T., L.P. Atkinson, W.B. Corlett, and J.L. Blanco, 2013. Gulf Stream's induced sea level rise and variability along the U.S. mid-Atlantic coast. *J. Geophys. Res.*, 118, 685–697.

Feddersen, F., 2014. The generation of surfzone eddies in a strong alongshore current. *J. Phys. Oceanogr.*, 44, 600–617.

Fewings, M., S.J. Lentz, and J. Fredericks, 2008. Observations of cross-shelf flow driven by cross-shelf wind on the inner continental shelf. *J. Phys. Oceanogr.*, 38, 2358–2378.

Fischer, A.S., R.A. Weller, D.L. Rudnick, C.C. Eriksen, C.M. Lee, K.H. Brink, C.A. Fox, and R.R. Leben, 2002. Mesoscale eddies, coastal upwelling, and the upper ocean heat budget in the Arabian Sea. *Deep-Sea Res. II*, 49, 2231–2264.

Flagg, C.N., and R.C. Beardsley, 1978. On the stability of the shelf water/slope water front south of New England. *J. Geophys. Res.*, 83, 4623–4631.

Fong, D.A., and W.R. Geyer, 2002. The alongshore transport of fresh water in a surface-trapped river plume. *J. Phys. Oceanogr.*, 32, 957–972.

Ford, W.L., J.R. Longard, and R.E. Banks, 1952. On the nature, occurrence and origin of cold low salinity water along the edge of the Gulf Stream. *J. Mar. Res.*, 11, 281–293.

Foreman, M.G.G., R.A. Walters, R.F. Henry, C.P. Keller, and A.G. Dolling, 1995. A tidal model for eastern Juan de Fuca Strait and the southern Strait of Georgia. *J. Geophys. Res.*, 100, 721–740.

Franks, P.J.S., and C. Chen, 1996. Plankton production in tidal fronts: A model of Georges Bank in summer. *J. Mar. Res.*, 54, 631–651.

Freilich, M., and A. Mahadevan, 2021. Coherent pathways for subduction from the surface mixed layer at ocean fronts. *J. Geophys. Res.*, 126, e2020JC017042 https://doi.org/10.1029/2020JC017042.

Fu, L.L., and B. Holt, 1982. SEASAT views oceans and sea ice with synthetic aperture radar. NASA/JPL Publ. 81–120, California Institute of Technology, Pasadena.

Gangopadhyay, A., G. Gawarkiewicz, E.N.S. Silva, A.M. Silver, M. Monim, and J. Clark, 2020. A census of the Warm-Core Rings of the Gulf Stream: 1980–2017. *J. Geophys. Res.*, 125, e2019JC016033, https://doi.org/10.1029/2019JC016033.

Ganju, N.K., S.J. Lentz, A.R. Kirincich, and J.T. Farrar, 2011. Complex mean circulation over the inner shelf south of Martha's Vineyard revealed by observations and a high-resolution model. *J. Geophys. Res.*, 116, C10036, https://doi.org/10.1029/2011JC007035.

Garrett, C., 1972. Tidal resonance in the Bay of Fundy and Gulf of Maine. *Nature*, 238, 441–443.

Garrett, C., P. MacCready, and P. Rhines, 1993. Boundary mixing and arrested Ekman layers: Rotating stratified flow near a sloping boundary. *Annu. Rev. Fluid Mech.*, 25, 291–323.

Garvine, R.W., 2001. The impact of model configuration in studies of buoyant coastal discharge. *J. Mar. Res.*, 59, 193–225.

Garvine, R.W., K.-C. Wong, and G. Gawarkiewicz, 1989. Quantitative properties of shelfbreak eddies. *J. Geophys. Res.*, 94, 14,475–14,483.

Gaube, P., D.B. Chelton, R.M. Samelson, M.G. Schlax, and L.W. O'Neill, 2015. Satellite observations of mesoscale eddy induced Ekman pumping. *J. Phys. Oceanogr.*, 45, 104–132.

Gawarkiewicz, G., 2000. Effects of ambient stratification and shelfbreak topography on offshore transport of dense water on continental shelves. *J. Geophys. Res.*, 105, 3307–3324.

Gawarkiewicz, G., K.H. Brink, F. Bahr, R.C. Beardsley, M. Caruso, and J.F. Lynch, 2004. A large-amplitude meander of the shelfbreak front south of New England: Observations from the shelfbreak PRIMER experiment. *J. Geophys. Res.*, 109, C03006, https://doi.org/10.1029/2002JC001468.

Geyer, W.R., and P. MacCready, 2014. The estuarine circulation. *Annu. Rev. Fluid Mech.*, 46, 175–197.

Gill, A.E., 1982. *Atmosphere-Ocean Dynamics*. Academic Press, New York, 662pp.

Gill, A.E., and A.J. Clarke, 1974. Wind-induced upwelling, coastal currents and sea level changes. *Deep-Sea Res.*, 21, 325–345.

Gill, A.E., and E.H. Schumann, 1974. The generation of long shelf waves by the wind. *J. Phys. Oceanogr.*, 9, 975–991.

Grant, W.D., and O.S. Madsen, 1986. The continental-shelf bottom boundary layer. *Annu. Rev. Fluid Mech.*, 18, 265–305.

Gregg, M.C., 2021. *Ocean Mixing*. Cambridge University Press, Cambridge, U.K., 400 pp.

Gregg, M.C., E.A. D'Asaro, J.J. Riley, and E. Kunze, 2018. Mixing efficiency in the ocean. *Annu. Rev. Mar. Sci.*, 10, 443–473, https://doi.org/10.1146/annurev-marine-121916-063643.

Gula, J., M.J. Molemaker, and J.C. McWilliams, 2015. Gulf Stream dynamics along the southeastern U.S. seaboard. *J. Phys. Oceanogr.*, 45, 690–715.

Haine, T.W.N., and J. Marshall, 1998. Gravitational, symmetric and baroclinic instability of the ocean mixed layer. *J. Phys. Oceanogr.*, 28, 634–657.

Haly-Rosendahl, K., F. Feddersen, D.B. Clark, and R.T. Guza, 2015. Surfzone to inner-shelf exchange estimated from dye tracer balances. *J. Geophys. Res.*, 120, 6289–6308.

Han, L., H. Seim, J. Bane, R.E. Todd, and M. Muglia, 2021. A shelf water cascading event near Cape Hatteras. *J. Phys. Oceanogr.*, 51, 2021–2033.

Harden, B.E., F. Straneo, and D.A. Sutherland, 2014. Moored observations of synoptic and seasonal variability in the East Greenland Coastal Current. *J. Geophys. Res.: Oceans*, 119, 8838–8857, https://doi.org/10.1002/2014JC010134.

Hartline, B.K., 1980. Coastal upwelling: Physical factors feed fish. *Science*, 208, 38–40.

Hasselmann, K., 1970. Wave-driven inertial oscillations. *Geophys. Fluid Dyn.*, 1, 463–502.

Helfrich K.R., and W.K. Melville, 2006. Long nonlinear internal waves. *Annu. Rev. Fluid Mech.* 38, 395–425.

Hendershott, M., and W. Munk, 1970. Tides. *Annu. Rev. Fluid Mech.*, 2, 205–224.

Hetland, R.D., 2017. Suppression of baroclinic instabilities in buoyancy-driven flow over sloping bathymetry. *J. Phys. Oceanogr.*, 47, 49–68.

Hickey, B.M, 1984. The fluctuating longshore pressure gradient on the Pacific northwest shelf: A dynamical analysis. *J. Phys. Oceanogr.*, 14, 276–293.

Hickey, B.M., L.J. Pietrafesa, D.A. Jay, and W.C. Boicourt, 1998. The Columbia River plume study: Subtidal variability in the velocity and salinity fields. *J. Geophys. Res.*, 103, 10,339–10,368.

Hogg, N.G., 1980. Observations of internal Kelvin waves trapped round Bermuda. *J. Phys. Oceanogr.*, 10, 1353–1376.

Holladay, C.G., and J.J. O'Brien, 1975. Mesoscale variability of sea surface temperatures. *J. Phys. Oceanogr.*, 5, 761–772.

Holton, J.R., and G.J. Hakim, 2012. *An Introduction to Dynamic Meteorology*, 5th ed. Academic Press, Waltham, Mass., 552pp.

Hong, B.G., W. Sturges, and A.J. Clarke, 2000. Sea level on the U.S. east coast: Decadal variability caused by open ocean wind-curl forcing. *J. Phys. Oceanogr.*, 30, 2088–2098.

Horner-Devine, A.R., 2009. The bulge circulation in the Columbia River plume. *Cont. Shelf Res.*, 29, 234–251.

Horner-Devine, A.R., D.A. Fong, and S.G. Monismith, 2008. Evidence for the inherent unsteadiness of a river plume: Satellite observation of the Niagara River discharge. *Limnol. Oceanogr.*, 53, 2731–2737.

Horner-Devine, A.R., R.D. Hetland, and D.G. MacDonald, 2015. Mixing and transport in coastal river plumes. *Annu. Rev. Fluid Mech.*, 47, 569–594.

Horton, B.P., R.E. Kopp, A.J. Garner, C.C. Hay, N.S. Khan, K. Roy, and T.A. Shaw, 2018. Mapping sea-level change in time, space, and probability. *Annu. Rev. Environ. Resour.*, 43, 481–521, https://doi.org/10.1146/annurev-environ-102017-025826.

Horwitz, R.M., and S.J. Lentz, 2016. The effect of wind direction on cross-shelf transport on an initially stratified inner shelf. *J. Mar. Res.*, 74, 201–227.

Howard, L.N., 1961. Note on a paper of John W. Miles. *J. Fluid Mech.*, 10, 509–512. https://doi.org/10.1016/B978-0-12-814003-1.00032-0.

Huthnance, J.M., 1973. Tidal current asymmetries over the Norfolk Sandbanks. *Estuar. Coast. Mar. Sci.*, 1, 89–99.

Huthnance, J.M., 1975. On trapped waves over a continental shelf. *J. Fluid Mech.*, 69, 689–704.

Huthnance, J.M., 1978. On coastal trapped waves: Analysis and numerical calculation by inverse iteration. *J. Phys. Oceanogr.*, 8, 74–92.

Huthnance, J.M., 2004. Ocean-to-shelf signal transmission: A parameter study. *J. Geophys. Res.*, 109, C12029, https://doi.org/10.1029/2004JC002358.

Huyer, A., J.H. Fleischbein, J. Kelser, P.M. Kosro, N. Perlin, R.L. Smith, and P.A. Wheeler, 2005, Two coastal upwelling domains in the northern California Current system. *J. Mar. Res.*, 63, 901–929.

Huyer, A., R.L. Smith, and T. Paluszkiewicz, 1987. Coastal upwelling off Peru during normal and El Niño times, 1981–1984. *J. Geophys. Res.*, 92, 14297–14,307.

Illig, S., M.-L. Bechèlery, and E. Cadier, 2018. Subseasonal coastal-trapped wave propagations in the southeastern Pacific and Atlantic Oceans: 2. Wave characteristics and connection with equatorial variability. *J. Geophys. Res.*,123, 3942–3961, https://doi.org/10.1029/ 2017JC013540.

Imbol Koungue, R.A., and P. Brandt, 2021. Impact of intraseasonal waves on Angolan warm and cold events. *J. Geophys. Res.*, 126, e2020JC017088. https://doi.org/10.1029/2020JC017088.

Ivanov, V.V., G.I. Shapiro, J.M. Huthnance, D.I. Aleynik, and P.N. Golovin, 2004. Cascades of dense water around the world ocean. *Prog. Oceanogr.*, 60, 47–98.

Jackson, R.H., S.J. Lentz, and F. Straneo, 2018. The dynamics of shelf forcing in Greenlandic fjords. *J. Phys. Oceanogr.*, 48, 2799–2827.

Janowitz, G.S., and L.J. Pietrafesa, 1980. A model and observation of time-dependent upwelling over the mid-shelf and slope. *J. Phys. Oceanogr.*, 10, 1574–1583.

Jiang, L., and R.W. Garwood Jr., 1996. Three-dimensional simulations of overflows on continental slopes. *J. Phys. Oceanogr.*, 26, 1214–1233.

Jithin, A.K., and P.A. Francis, 2021. Formation of an intrathermocline eddy triggered by the coastal-trapped wave in the northern Bay of Bengal. *J. Geophys. Res.*, 126, e2021JC017725. https://doi.org/10.1029 /2021JC017725.

Johannessen, J.A., E. Svendsen, S. Sandven, O.M. Johannessen, and K. Lygre, 1989. Three-dimensional structure of mesoscale eddies in the Norwegian Coastal Current. *J. Phys. Oceanogr.*, 19, 3–19.

Johnson, E.R., 1991. The scattering at low frequencies of coastally trapped waves. *J. Phys. Oceanogr.*, 21, 913–832.

Kadko, D.C., L. Washburn, and B. Jones, 1991. Evidence of subduction within cold filaments of the northern California coastal transition zone. *J. Geophys. Res.*, 96, 14,909–14,926.

Kelly, K.A., and D.C. Chapman, 1988. The response of stratified shelf and slope waters to steady offshore forcing. *J. Phys. Oceanogr.*, 18, 906–925.

Kelly, S.M., 2019. Coastally generated near-inertial waves. *J. Phys. Oceanogr.*, 49, 2979–2995.

Kelly, S.M., and S. Ogbuka, 2022. Coastal trapped waves: Normal modes, evolution equations, and topographic generation. *J. Phys. Oceanogr.*, 52, 1835–1848.

Kirincich, A., 2016. The occurrence, drivers, and implications of submesoscale eddies on the Martha's Vineyard inner shelf. *J. Phys. Oceanogr.*, 46, 2645–2662.

Kirincich, A., B. Hodges, P. Flament, and V. Frisch, 2022. Horizontal stirring over the northeast U.S. continental shelf: The spatial and temporal evolution of surface eddy kinetic energy. *J. Geophys. Res.*, 127, https://doi.org/10.1029/2021JC017307.

Kirincich, A., S.J. Lentz, J.T. Farrar, and N.K. Ganju, 2013. The spatial structure of tidal and mean circulation over the inner shelf south of Martha's Vineyard, Massachusetts. *J. Phys. Oceanogr.*, 43, 1940–1958.

Komar, P.D., 1998. *Beach Processes and Sedimentation*, 2nd ed. Prentice Hall, Hoboken, N.J., 544 pp.

Kopp, R.E., C.C. Hay, C.M. Little, and J.X. Mitrovica, 2015. Geographic variability of sea level change. *Curr. Clim. Change Rep.* 1, 192–204, https://doi.org/10.1007/s40641-015-0015-5.

Kumar, N., and F. Feddersen, 2017: The effects of Stokes drift and transient rip currents on the inner shelf. Part II: With stratification. *J. Phys. Oceanogr.*, 47, 243–260, https://doi.org/10.1175/JPO-D-16-0077.

Krug, M., S. Swart, and J. Gula, 2017. Submesoscale cyclones in the Agulhas current. *Geophys. Res. Lett.*, 44, 346–354.

Kundu, P.K., and J.S. Allen, 1976. Some three-dimensional characteristics of low-frequency current fluctuations near the Oregon coast. *J. Phys. Oceanogr.*, 6, 181–199, https://doi.org/10.1175/1520 -0485(1976)006<0181:STDCOL>2.0.CO;2.

Kundu, P.K., S.-Y. Chao, and J.P. McCreary, 1983. Transient coastal currents and inertia-gravity waves. *Deep-Sea Res.*, 30, 1059–1082.

LaCasce, J.H., 2008. Statistics from Lagrangian observations. *Prog. Oceanogr.*, 77, 1–29.

Lamb, H., 1932. *Hydrodynamics*, 6th ed. Cambridge University Press, Cambridge, U.K., 738pp.

Lamb, K.G., 2014. Internal wave breaking and dissipation mechanisms on the continental slope/shelf. *Annu. Rev. Fluid Mech.*, 46, 231–254.

Lane, E.M., J.M. Restrepo, and J.C. McWilliams, 2007. Wave-current interaction: A comparison of radiation-stress and vortex-force representations. *J. Phys. Oceanogr.*, 37, 1122–1141.

LeBlond, P.H., and L.A. Mysak, 1978. *Waves in the Ocean*. Elsevier, New York, 602 pp.

Lee, C.M., and K.H. Brink, 2010. Observations of storm-induced mixing and Gulf Stream Ring incursion over the southern flank of Georges Bank: Winter and summer 1997. *J. Geophys. Res.*, 115, C08008, https://doi.org/10.1029/2009JC005706.

Lee, T.N., W.J. Ho, V. Kourafalou, and J.D. Wang, 1984. Circulation on the continental shelf of the southeastern United States. Part I: Subtidal response to wind and Gulf Stream forcing during winter. *J. Phys. Oceanogr.*, 14, 1001–1012.

Lee, T.N., J.A. Yoder, and L.P. Atkinson, 1991. Gulf Stream frontal eddy influence on productivity of the southeast U.S. continental shelf. *J. Geophys. Res.*, 96, 22,191–22,205.

Legeckis, R.V., 1979. Satellite observations of the influence of bottom topography on the seaward deflection of the Gulf Stream off Charleston, South Carolina. *J. Phys. Oceanogr.*, 9, 483–497.

Lentz, S.J., 1992. The surface boundary layer in coastal upwelling regions. *J. Phys. Oceanogr.*, 22, 1517–1539.

Lentz, S.J., 1995. Sensitivity of the inner shelf circulation to the form of the eddy viscosity profile. *J. Phys. Oceanogr.*, 25, 19–28.

Lentz, S.J., 2004. The response of buoyant coastal plumes to upwelling-favorable winds. *J. Phys. Oceanogr.*, 34, 2458–2469.

Lentz, S.J., 2010. The mean along-isobath heat and salt balance over the Middle Atlantic Bight continental shelf. *J. Phys. Oceanogr.*, 40, 934–948.

Lentz, S.J., B. Butman, and C. Harris, 2014, The vertical structure of the circulation and dynamics in Hudson Shelf Valley. *J. Geophys. Res.*, 119, 3694–3713.

Lentz, S.J., and D.C. Chapman, 2004, The importance of cross-shelf nonlinear momentum flux during wind-driven coastal upwelling. *J. Phys. Oceanogr.*, 34, 2444–2457.

Lentz, S.J., K.A. Davis, J.H. Churchill, and T.M. DeCarlo, 2017. Coral reef drag coefficients—water depth dependence. *J. Phys. Oceanogr.*, 47, 1061–1075.

Lentz, S.J., and M.R. Fewings, 2012. The wind- and wave-driven inner shelf circulation. *Annu. Rev. Mar. Sci.*, 4, 317–343.

Lentz, S.J., M.R. Fewings, P. Howd, J. Fredericks, and K Hathaway, 2008. Observations and a model of undertow over the inner continental shelf. *J. Phys. Oceanogr.*, 38, 2341–2357.

Lentz, S.J., R.T. Guza, S. Elgar, F. Feddersen, and T.H.C. Herbers, 1999. Momentum balances on the North Carolina inner shelf. *J. Geophys. Res.*, 104, 18,205–18,226.

Lentz, S.J., and K.R. Helfrich, 2002. Buoyant gravity currents along a sloping bottom in a rotating frame. *J. Fluid Mech.*, 464, 251–278.

Lentz, S.J., and B. Raubenheimer, 1999. Field observations of wave setup. *J. Geophys. Res.*, 104, 25,867–25,875.

Lentz, S.J., and J.H. Trowbridge, 1991. The bottom boundary layer over the northern California shelf. *J. Phys. Oceanogr.*, 21, 1186–1201.

Lentz, S.J., and J.H. Trowbridge, 2001. A dynamical description of fall and winter mean current profiles over the northern California shelf. *J. Phys. Oceanogr.*, 31, 914–931.

Lerczak, J.A., and W.R. Geyer, 2004. Modeling the lateral circulation in straight, stratified estuaries. *J. Phys. Oceanogr.*, 34, 1410–1428.

Li, X., W. Zhang, and Z. Rong, 2021. The interaction between warm-core rings and submarine canyons and its influence on the onshore transport of offshore waters. *J. Geophys. Res. Oceans*, 126, e2021JC017989. https://doi. org/10.1029/2021JC017989.

Linder, C.A., and G. Gawarkiewicz, 1998. A climatology of the shelfbreak front in the Middle Atlantic Bight. *J. Geophys. Res.*, 103, 18,405–18,423.

Loder, J.W., 1980. Topographic rectification of tidal currents on the sides of Georges Bank. *J. Phys. Oceanogr.*, 10, 1399–1416.

Loder, J.W., D. Brickman, and E.P.W. Horne, 1992. Detailed structure of currents and hydrography on the northern side of Georges Bank. *J. Geophys. Res.*, 97 (C9), 14,331–14,351, https://doi.org/10.1029/92JC01342.

Loder, J.W., and D.A. Greenberg, 1986. Predicted positions of tidal fronts in the Gulf of Maine region. *Cont. Shelf Res.*, 6, 397–414.

Longuet-Higgins, M.S., 1953. Mass transport in water waves. *Philos. Trans. Royal Soc. A*, 245A, 535–581.

Longuet-Higgins, M.S., and R.W. Stewart, 1962. Radiation stress and mass transport in gravity waves, with application to "surf beat." *J. Fluid Mech.*, 13, 481–504.

Longuet-Higgins, M.S., and R.W. Stewart, 1964. Radiation stress in water waves: A physical discussion, with applications. *Deep-Sea Res.*, 11, 529–562.

López-Mariscal, M., and A.J. Clarke, 1993. On the influence of wind-stress curl on low-frequency shelf water flow. *J. Phys. Oceanogr.*, 23, 2717–2727.

Louis, J.P., and P.C. Smith, 1982. The development of the barotropic radiation field of an eddy over a slope. *J. Phys. Oceanogr.*, 12, 56–73.

Luther, M.E., and J.M. Bane, 1985. Mixed instabilities in the Gulf Stream over the continental slope. *J. Phys. Oceanogr.*, 15, 3–23.

Lynch J.F., S.R. Ramp, C.-S. Chin, T.Y. Tang, Y.-J. Yang, and J.A. Simmen, 2004. Research highlights from the Asian Seas International Acoustics Experiment in the South China Sea. *IEEE J. Ocean. Eng.*, 29, 1067–74.

Maas, L.R.M., and J.T.F. Zimmerman, 1989. Tide-topography interactions in a stratified shelf sea II. Bottom trapped internal tides and baroclinic residual currents. *Geophys. Astrophys. Fluid Dyn.*, 45, 37–69.

MacCready, P., and P.B. Rhines, 1991. Buoyant inhibition of Ekman transport on a slope and its effect on stratified spin-up. *J. Fluid Mech.*, 223, 631–661.

MacCready, P., and W.R. Geyer, 2009. Advances in estuarine physics. *Annu. Rev. Marine Sci.*, 2, 35–58.

MacDonald, D.G., L. Goodman, and R.D. Hetland, 2007. Turbulent dissipation in a near-field river plume: A comparison of control volume and microstructure observations with a numerical model. *J. Geophys. Res.*, 112:C07026.

Mackas, D.L., P.T. Strub, A. Thomas, and V. Montecino, 2005. "Eastern Ocean Boundaries Pan-regional Overview (E)." In *The Sea*, vol. 14, A.R. Robinson and K.H. Brink (eds.). Harvard University Press, Cambridge, Mass., 21–59.

Madsen, O.S. 1977. A realistic model of the wind-induced Ekman boundary layer. *J. Phys. Oceanogr.*, 7, 248–55.

Malan, N., M. Archer, M. Roughan, P. Cetina-Heredia, M. Hemming, C. Rocha, A. Schaeffer, I. Suthers, and E. Queiroz, 2020. Eddy-driven cross-shelf transport in the East Australian Current Separation Zone. *J. Geophys. Res.*, 125, e2019JC015613. https://doi.org/10.1029/2019JC015613.

Masse, A.K. and C.R. Murthy, 1990. Observations of the Niagara River thermal plume (Lake Ontario, North America). *J. Geophys. Res.*, 95, 16,097–16,109.

McCreary, J.P., Jr. and S.-Y. Chao, 1985. Three-dimensional shelf circulation along an eastern boundary. *J. Mar. Res.*, 43, 13–36.

McCreary, J.P., H.S. Lee, and D.B. Enfield, 1989. The response of the coastal ocean to strong offshore winds: With application to circulations in the Gulfs of Tehuantepec and Papagayo. *J. Mar. Res.*, 47, 81–109.

McDowell, S.E, and H.T. Rossby, 1978. Mediterranean water: An intense eddy off the Bahamas. *Science*, 202, 1085–1087.

McPhee-Shaw, E.E., D.A. Siegel, L. Washburn, M.A. Brzezinski, J.L. Jones, A. Leydecker, and J. Melack, 2007. Mechanisms for nutrient delivery to the inner shelf: Observations from the Santa Barbara Channel. *Limnol. Oceanogr.*, 52, 1748–1766.

McSweeney, J.M., M.R. Fewings, J.A. Lerczak, and J.A. Barth, 2021. The evolution of a northward-propagating buoyant coastal plume after a wind relaxation event. *J. Geophys. Res.*, 126, e2021JC017720, https://doi.org/10.1029/2021JC017720.

McSweeney, J.M., J.A. Lerczak, J.A. Barth, J. Becherer, J.A. MacKinnon, A.F. Waterhouse, J.A. Colosi, J.H. MacMahan, F. Fedderson, J. Calantoni, A. Simpson, S. Celona, M.C. Haller, and E. Terrill, 2020. Alongshore variability of shoaling internal bores on the inner shelf. *J. Phys. Oceanogr.*, 50, 2965–2981.

McWilliams, J.C., J.M. Restrepo and E.M. Lane, 2004. An asymptotic theory for the interaction of waves and currents in coastal waters. *J. Fluid Mech.*, 511, 135–178.

Mei, C.C., 1983. *The Applied Dynamics of Ocean Surface Waves*. Wiley-Interscience, New York, 740pp.

Middleton, J.F., and D. Ramsden, 1996, The evolution of the bottom boundary layer on the sloping continental shelf: A numerical study. *J. Geophys. Res.*, 101 (C8), 18,061–18,077.

Miles, J.W., 1972. Kelvin waves on oceanic boundaries. *J. Fluid Mech.*, 55, 113–127.

Mitchum, G.T., and A.J. Clarke, 1986a. The frictional nearshore response to forcing by synoptic scale winds. *J. Phys. Oceanogr.*, 16, 934–946.

Mitchum, G.T., and A.J. Clarke, 1986b. Evaluation of frictional, wind-forced, long-wave theory on the west Florida shelf. *J. Phys. Oceanogr.*, 16, 1029–1037.

Mitrovica, J.X., M.E, Tamisiea, J.L. Davis, and G.L. Milne, 2001. Recent mass balance of polar ice sheets inferred from patterns of global sea-level change. *Nature*, 409, 1026–1029.

Moffat, C., and S. Lentz, 2014. On the response of a buoyant plume to downwelling-favorable wind stress. *J. Phys. Oceanogr.*, 42, 1083–1098.

Moody, J.A., R.C. Beardsley, W.S. Brown, P. Daifuku, D.A. Mayer, H.O. Mofjeld, B. Petrie, S. Ramp, P. Smith, and W.R Wright, 1984. Atlas of tidal elevation and current observations on the northeast American continental shelf and slope. *U.S. Geolog. Surv. Bull.* 1611, U.S. GPO, 122pp.

Mooers, C.N.K., 1975. Several effects of baroclinic currents on the three-dimensional propagation of inertial-internal waves. *Geophys. Fluid Dyn.*, 6, 227–284.

Mooers, C.N.K., C.A. Collins, and R.L. Smith, 1976. The dynamic structure of the frontal zone in the coastal upwelling region off Oregon. *J. Phys. Oceanogr.*, 6, 3–21.

Moore, D.W., 1968. Planetary-gravity waves in an equatorial ocean. PhD diss., Harvard University, Cambridge, Mass., 207 pp.

Moulton M., C.C. Chickadel, and J. Thomson, 2021. Warm and cool nearshore plumes connecting the surf zone and the inner shelf. *Geophys. Res. Lett.*, 48, e2020GL091675.

Moum, J.N., 2021. Variations in ocean mixing from seconds to years. *Annu. Rev. Mar. Sci.*, 13, https://doi.org/10.1146/annurev-marine-031920-122846.

Mountain, D.G., M. Pastuszak, and D.A. Busch, 1989. Slope water intrusion to the Great South Channel during autumn, 1977–1985. *J. Northwest Atl. Fish. Sci.*, 9, 97–102.

Münchow, A., and R.W. Garvine, 1993. Buoyancy and wind forcing of a coastal current. *J. Mar. Res.*, 51, 293–322.

Münchow A., A.K. Masse, and R.W. Garvine, 1992. Astronomical and nonlinear tidal currents in a coupled estuary/shelf system. *Cont. Shelf Res.*, 12, 471–498.

Nash, J.D., S.M. Kelly, E.L. Shroyer, J.N. Moum, and T.F. Duda, 2012. The unpredictable nature of internal tides on continental shelves. *J. Phys. Oceanogr.*, 42, 1981–2000.

National Research Council, 2012. *Sea-Level Rise for the Coasts of California, Oregon, and Washington: Past, Present, and Future.* Washington, DC: The National Academies Press. 216pp, https://doi.org/10.17226/13389.

Newton, C.W., 1978. Fronts and wave disturbances in Gulf Stream and atmospheric jet stream. *J. Geophys. Res.*, 83, 4697–4706.

Nicholls, R.J., 2007. The impacts of sea level rise. *Ocean Challenge,* 15, 13–17.

Niiler, P.P., 1969. On the Ekman divergence in an oceanic jet. *J. Geophys. Res.*, 74, 7048–7052, https://doi.org/10.1029/JC074i028p07048.

Niiler, P.P., 1975. Deepening of the wind-mixed layer. *J. Mar. Res.*, 33, 405–422.

Nof, D., and T. Pichevin, 2001. The ballooning of outflows. *J. Phys. Oceanogr.*, 31, 3045–3058.

Nof, D., T. Pichevin, and J. Sprintall, 2002. "Teddies" and the origin of the Leeuwin Current. *J. Phys. Oceanogr.*, 32, 2571–2588.

Okubo, A., 1971. Oceanic diffusion diagrams. *Deep-Sea Res.*, 18, 789–802.

Oltman-Shay, J., and R.T. Guza, 1987. Infragravity edge wave observations on two California beaches. *J. Phys. Oceanogr.*, 17, 644–663.

Oltman-Shay, J., P.A. Howd, and W.A. Birkemeier, 1989. Shear instabilities of the mean longshore current: 2, Field data. *J. Geophys Res.*, 94, 18,031–18,042.

Palomares, M.-L.D. and D. Pauly, 2019. "Coastal Fisheries: Past, Present, and Possible Futures." In *Coasts and Estuaries: The Future*, E. Wolanski, J.W. Day, M. Elliott, and R. Ramachandran. (eds.). Elsevier, Amsterdam, 569–576.

Pearson, B., 2018. Turbulence-induced anti-Stokes flow and the resulting limitations of large-eddy simulations. *J. Phys. Oceanogr.*, 48, 117–122.

Pedlosky, J., 1978. A nonlinear model of the onset of upwelling. *J. Phys. Oceanogr.*, 8, 178–187.

Pedlosky, J., 1979. *Geophysical Fluid Dynamics.* Springer-Verlag, New York, 624pp.

Pedlosky, J., 2003. *Waves in the Ocean and Atmosphere.* Springer-Verlag, New York, 272pp.

Perlin, A., J.N. Moum, J.M. Klymak, M.D. Levine, T. Boyd, and P.M. Kosro, 2005. A modified law-of-the-wall applied to oceanic bottom boundary layers. *J. Geophys. Res.*, 110, C10S10, https://doi.org/10.1029/2004JC002310.

Pettigrew, N.R., 1980. The dynamics and kinematics of the coastal boundary layer off Long Island. PhD thesis, Massachusetts Institute of Technology, Cambridge, Mass., 262 pp., https://doi.org/10.1575/1912/ 3727.

Pickart, R.S., 1995. Gulf Stream–generated topographic Rossby waves. *J. Phys. Oceanogr.*, 25, 574–586.

Pinardi, N., P.F.J. Lermusiaux, K.H. Brink, and R. Preller, eds., 2017. *The Sea*. Volume 17, *The Science of Ocean Prediction*. Print Supplement, *J. Mar. Res.*, 75.

Pingree, R.D., and D.K. Griffiths, 1978. Tidal fronts on the shelf seas around the British Isles. *J. Geophys. Res.*, 83, 4615–4622.

Piola, A.R., O.O. Möller Jr., R.A. Guerrero, and E.J.D. Campos, 2008. Variability of the subtropical front off eastern South America: Winter and summer 2004. *Cont. Shelf Res.*, 28, 1639–1648.

Platzman, G.W., G.A. Curtis, K.S. Hansen, and R.D. Slater, 1981. Normal modes of the world ocean. Part II: Description of modes in the period range of 8 to 80 hours. *J. Phys. Oceanogr.*, 11, 579–603.

Pollard, R.T., P.B. Rhines, and R.O.R.Y. Thompson, 1973. The deepening of the wind-mixed layer. *Geophys. Fluid Dyn.*, 3, 381–404.

Ponte, R.M., B. Meyssignac, C.M. Domingues, D. Stammer, A. Cazenave, and T. Lopez (eds.), 2020. *Relationship Between Coastal Sea Level and Large Scale Ocean Circulation*. Springer, Cham, Switzerland, 450pp.

Pörtner, H.-O., D.C. Roberts, V. Masson-Delmotte, P. Zhai, M. Tignor, E. Poloczanska, K. Mintenbeck, A. Alegría, M. Nicolai, A. Okem, et al. (eds.), 2019. *IPCC Special Report on the Ocean and Cryosphere in a Changing Climate*. Intergovernmental Panel on Climate Change, Switzerland, 477–587.

Power, S.B., J.H. Middleton, and R.H.J. Grimshaw, 1989. Frictionally modified continental shelf waves and the subinertial response to wind and deep-ocean forcing. *J. Phys. Oceanogr.*, 19, 1486–1506.

Pratt, L.J., and J.A. Whitehead, 2007. *Rotating Hydraulics: Nonlinear Topographic Effects in the Ocean and Atmosphere*. Springer, New York, 592pp.

Price, J.F., R.A. Weller, and R. Pinkel, 1986. Diurnal cycling: Observations and models of the upper ocean response to diurnal heating, cooling, and wind mixing. *J. Geophys. Res.*, 91, 8411–8427.

Pringle, J.M., 2001. Cross-shelf eddy heat transport in a wind-free coastal ocean during winter time cooling. *J. Geophys. Res.*, 106, 2589–2604.

Pringle, J.M., and E.P. Dever, 2009. Dynamics of wind-driven upwelling and relaxation between Monterey Bay and Point Arena: Local-, regional, and gyre-scale controls. *J. Geophys. Res.*, 114, C07003, https://doi.org/10.1029/2008JC005016.

Pritchard, M., and R.A. Weller, 2005. Observations of internal bores and waves of elevation on the New England inner continental shelf during summer 2001. *J. Geophys. Res.*, 110, C03020, https://doi.org/10.1029/2004JC002377.

Rabalais, N.N., W,-J. Cai, J. Carstensen, D.J. Conley, B. Fry, X. Hu, Z. Quinones-Rivera, R. Rosenberg, C.P. Slomp, R.E. Turner, M. Voss, B. Wissel, and J. Zhang, 2014. Eutrophication-driven deoxygenation in the coastal ocean. *Oceanography*, 27, 172–183.

Rabalais, N.N., R.E. Turner, R.J. Díaz, and D. Justić, 2009. Global change and eutrophication of coastal waters. *ICES J. Mar. Sci.*, 66, 1528–1537.

Ray, R.D., 1999. A global ocean tide model from TOPEX/POSEIDON altimetry: GOT99.2. NASA technical report NASA/TM-1999-209478, 58pp.

Reid, R.O., 1958. Effect of Coriolis force on edge waves (I) Investigation of the normal modes. *J. Mar. Res.*, 16, 109–144.

Reniers, A.J.H.M., E.B. Thornton, T.P. Stanton, and J.A. Roelvink, 2004. Vertical flow structure during Sandy Duck: Observations and modeling. *Coast. Eng.*, 51, 237–260.

Rennie, S.E., J.L. Largier, and S.J. Lentz, 1999. Observations of a pulsed buoyancy current downstream of Chesapeake Bay. *J. Geophys. Res.*, 104, 18,227–18,240.

Rhines, P.B., 1977. "The Dynamics of Unsteady Currents." In *The Sea*, vol. 6, E.D. Goldberg, I.N. McCave, J.J. O'Brien, and J.H. Steele (eds.). John Wiley & Sons, New York, 189–318.

Robinson, A.R., 1964. Continental shelf waves and the response of sea level to weather systems. *J. Geophys. Res.*, 69, 367–368.

Robinson, A.R., and K.H. Brink (eds.), 1998. *The Global Coastal Ocean: Regional Studies and Syntheses*, vol.11, *The Sea*. John Wiley & Sons, New York, 1079pp.

Robinson, A.R., and K.H. Brink (eds.), 2006. *The Global Coastal Ocean: Interdisciplinary Regional Studies and Syntheses*, vol.14, *The Sea*. Harvard University Press, Boston, Mass., 1535pp.

Rodriguez, A.R, S.N. Giddings, and N. Kumar, 2018. Impacts of nearshore wave-current interaction on transport and mixing of small-scale buoyant plumes. *Geophys. Res. Lett.*, 45, 8379–8389.

Rosenfeld, L.K., 1990. Baroclinic semidiurnal tidal currents over the continental shelf off northern California. *J. Geophys. Res.*, 95, 22,153–22,172.

Samelson, R.M., 2017. Time-dependent linear theory for the generation of poleward undercurrents on eastern boundaries. *J. Phys. Oceanogr.*, 47, 3037–3059.

Savidge, D.K., and J.A. Austin, 2007. The Hatteras front: August 2004 velocity and density structure. *J. Geophys. Res.*, 112, C07006, https://doi.org/10.1029/2006JC003933.

Savidge, D.K., and W.B. Savidge, 2014. Seasonal export of South Atlantic Bight and Mid-Atlantic Bight shelf waters at Cape Hatteras. *Cont. Shelf Res.*, 74, 50–59.

Schaeffer, A., A. Gramoulle, M. Roughan, and A. Mantovanelli, 2017. Characterizing frontal eddies along the East Australian Current from HF radar observations. *J. Geophys. Res.*, 122, 3964–3980.

Shapiro, G.I., J.M. Huthnance, and V.V. Ivanov, 2003. Dense water cascading off the continental shelf. *J. Geophys. Res.*, 108, 3390, https://doi.org/10.1029/2002JC001610.

Shearman, R.K., and K.H. Brink, 2010. Evaporative dense water formation and cross-shelf exchange over the northwest Australian inner shelf. *J. Geophys, Res.*, 115, C06027, https://doi.org/10.1029/2009JC005931.

Shi, C., and D. Nof, 1993. The splitting of eddies along boundaries. *J. Mar. Res.*, 51, 771–795.

Shroyer, E.L., J.N. Moum, and J.D. Nash, 2010. Energy transformations and dissipation of nonlinear internal waves over New Jersey's continental shelf. *Nonlin. Processes Geophys.*, 17, 345–360.

Simpson, J.H., 1997. Physical processes in the ROFI regime. *J. Mar. Syst.*, 12, 3–15.

Simpson J.H., and J.R. Hunter, 1974. Fronts in the Irish Sea. *Nature*, 250, 404–406.

Simpson, J.H., and I.D. James, 1986. "Coastal and Estuarine Fronts." In *Baroclinic Processes on Continental Shelves*, C.N. K. Mooers (ed.). American Geophysical Union, Washington, DC, 63–94.

Simpson, J.H., and J. Sharples, 2012. *Physical and Biological Oceanography of Shelf Seas*. Cambridge University Press, Cambridge, U.K., 448 pp.

Slinn, D.N., J.S. Allen, P.A. Newberger, and R.A. Holman, 1998. Nonlinear shear instabilities of alongshore currents over barred beaches. *J. Geophys. Res.*, 103,18,357–18,379.

Smith, R., 1972. Nonlinear Kelvin and continental-shelf waves. *J. Fluid Mech.*, 52, 379–391.

Smith, R.L., 1968. Upwelling. *Oceanogr. Mar. Biol. Annu. Rev.*, 6, 11–46.

Smith, R.L., 1978. Poleward propagating perturbations in current and sea levels along the Peru coast. *J. Geophys. Res.*, 83, 6083–6092.

Smith, R.L., 1981. "A Comparison of the Structure and Variability of the Flow Field in Three Coastal Upwelling Regions: Oregon, Northwest Africa, and Peru." In *Coastal Upwelling* F.A. Richards (ed.). American Geophysical Union, Washington, D.C., 107–118.

Stammer, D., A. Cazenave, R.M. Ponte, and M.E. Tamisiea, 2013. Causes for contemporary regional sea level changes. *Annu. Rev. Mar. Sci.*, 5, 21–46.

Stern, M.E., 1965. Interaction of a uniform wind stress with a geostrophic vortex. *Deep-Sea Res. Oceanogr. Abstr.*, 12, 355–367, https://doi.org/10.1016/0011-7471(65)90007-0.

Stevenson, M.R., R.W. Garvine, and B. Wyatt, 1974. Lagrangian measurements in a coastal upwelling zone off Oregon. *J. Phys. Oceanogr.*, 4, 321–336.

Stokes, G.G., 1847. On the theory of oscillatory waves. *Trans. Camb. Philos. Soc.*, 8, 441–455.

Stommel, H.M., and A. Leetmaa, 1972. Circulation on the continental shelf. *Proc. Nat. Ac. Sci.*, 69, 3380–3384.

Stone, P.H., 1966. On non-geostrophic baroclinic instability. *J. Atmos. Sci.*, 23, 390–400.

Straneo, F., and C. Cenedese, 2015. The dynamics of Greenland's glacial fjords and their role in climate. *Annu. Rev. Mar. Sci.*, 7, 89–112.

Strub, P.T., P.M. Kosro, A. Huyer, and CTZ collaborators, 1991. The nature of cold filaments in the California Current system. *J Geophys. Res.*, 96, 14,743–14,768.

Suanda, S.H., N. Kumar, A.J. Miller, E. Di Lorenzo, K. Haas, D. Cai, C.A. Edwards, L. Washburn, M.R. Fewings, R. Torres, and F. Feddersen, 2016. Wind relaxation and a coastal buoyant plume north of Pt. Conception. *J. Geophys. Res.*, 121, 7455–7475, https://doi.org/10.1002/2016JC011919.

Sullivan, P.P., and J.C. McWilliams, 2010. Dynamics of winds and currents coupled to surface waves. *Annu. Rev. Fluid Mech.*, 42, 19–42.

Talley, L.D., G.L. Pickard, W.J. Emery, and J.H. Swift, 2011. *Descriptive Physical Oceanography: An Introduction.* Academic Press, London, 564pp.

Thompson, R.O.R.Y., 1973. Stratified Ekman boundary layer models. *Geophys. Fluid. Dyn.*, 5, 201–210.

Thomson, R.E., and W.J. Emery, 2014. *Data Analysis Methods in Physical Oceanography.* Elsevier Science, Waltham, Mass., 728pp.

Thorade, H., 1909. Über die kalifornische Meeresströmung, Oberflächentemperaturen und Strömungen an der Westküste Nordamerikas. *Ann. der Hydrog. Mar. u. Meteor., Berlin*, 37, 17–34, 63–76.

Thornton, E.B., and R.T. Guza, 1986. Surf zone longshore currents and random waves: Field data and models. *J. Phys. Oceanogr.*, 16, 1165–1178.

Thorpe, S.A., 2004. Langmuir circulation. *Annu. Rev. Fluid Mech.*, 36,55–79.

Thorpe, S.A., 2007. *An Introduction to Ocean Turbulence.* Cambridge University Press, Cambridge, U.K., 242pp.

Timko, P.G., B.K. Arbic, P. Hyder, J.G. Richman, L. Zamudio, E. O'Dea, A.J. Wallcraft, and J.F. Shriver, 2019. Assessment of shelf sea tides and tidal mixing fronts in a global ocean model. *Ocean Modelling*, 136, 66–84.

Todd, R.E., 2020. Export of Middle Atlantic Bight shelf waters near Cape Hatteras from two years of underwater glider observations. *J. Geophys. Res.*, 125, e2019JC016006.

Trasviña, A., E.D. Barton, J. Brown, H.S. Velez, P.M. Kosro, and R.L. Smith, 1995. Offshore wind forcing in the Gulf of Tehuantepec, Mexico: The asymmetric circulation. *J. Geophys. Res.*, 100, 20,649–20,663.

Trowbridge, J.H., and S.J Lentz, 1991. Asymmetric behavior of an oceanic boundary layer above a sloping bottom. *J. Phys. Oceanogr.*, 21, 1171–1185.

Trowbridge, J.H., and S.J. Lentz, 1998. Dynamics of the bottom boundary layer on the northern California shelf. *J. Phys. Oceanogr.*, 28, 2075–2093.

Trowbridge, J.H., and S.J. Lentz, 2018. The bottom boundary layer. *Annu. Rev. Mar. Sci.*, 10, 397–420.

Turner, J.S., 1973. *Buoyancy Effects in Fluids.* Cambridge University Press, Cambridge, U.K., 367pp.

Uchyama, Y., J.C. McWilliams, and A.F. Shchepetkin, 2010. Wave-current interaction in an oceanic circulation model with a vortex-force formalism: Application to the surf zone. *Ocean Dyn.*, 34, 16–35.

Valle-Levinson, A., K.-C. Wong, and K.T. Bosley, 2001. Observations of the wind-induced exchange at the entrance to Chesapeake Bay. *J. Mar. Res.*, 59, 391–416.

Vallis, G.K., 2017. *Atmospheric and Oceanic Fluid Dynamics: Fundamentals and Large-Scale Circulation.*, 2nd ed. Cambridge University Press, Cambridge, U.K., 964pp.

Velez-Belchi, P., L.R. Centurioni, D.-K. Lee, S. Jan, and P.P. Niiler, 2013. Eddy-induced Kuroshio intrusions onto the continental shelf of the East China Sea. *J. Mar. Res.*, 71, 83–108.

Vlasenko, V., N. Stashchuk, and K. Hutter, 2005. *Baroclinic Tides.* Cambridge University Press, Cambridge, U.K., 351 pp.

von Appen, W.-J., R.S. Pickart, K.H. Brink, and T.W. Haine, 2014. Water column structure and statistics of Denmark Strait overflow water cyclones. *Deep-Sea Res. I*, 84, 110–126.

Wåhlin, A.K., 2002. Topographic steering of dense currents with application to submarine canyons. *Deep-Sea Res. I*, 49, 305–320.

Wang, D.-P., and A.J. Elliott, 1978. Non-tidal variability in the Chesapeake Bay and Potomac River: Evidence for non-local forcing. *J. Phys. Oceanogr.*, 8, 225–232.

Wang, P., J.C. Williams, Y. Uchyama, M.D. Chekroun, and D.L. Yi, 2020. Effects of wave streaming and wave variations on near-shore wave-driven circulation. *J. Phys. Oceanogr.*, 50, 3025–3041.

Warner, S.J., P. MacCready, J.N. Moum, and J.D. Nash, 2013. Measurement of tidal form drag using seafloor pressure sensors. *J. Phys. Oceanogr.*, 43, 1150–1172.

Washburn, L., M.R. Fewings, C. Melton, and C. Gotschalk, 2011. The propagating response of coastal circulation due to wind relaxations along the central California coast. *J. Geophys. Res.*, 116, C12028, https://doi.org/10.1029/2011JC007502.

Waterhouse, A.F., J.A. Mackinnon, R.C. Musgrave, S.M. Kelly, A. Pickering, and J. Nash, 2017. Internal tide convergence and mixing in a submarine canyon. *J. Phys. Oceanogr.*, 47, 303–322.

WCRP Global Sea Level Budget Group, 2018. Global sea-level budget 1993–present. *Earth Syst. Sci. Data*, 10, 1551–1590, https://doi.org/10.5194/essd-10-1551-2018.

Weatherly, G.L., and P.J. Martin, 1978. On the structure and dynamics of the oceanic bottom boundary layer. *J. Phys. Oceanogr.*, 8, 557–570.

Wenegrat, J.O., J. Callies, and L.N. Thomas, 2018. Submesoscale baroclinic instability in the bottom boundary layer. *J. Phys. Oceanogr.*, 48, 2571–2592.

Wenegrat, J.O., and L.N. Thomas, 2017. Ekman transport in balanced currents with curvature. *J. Phys. Oceanogr.*, 47, 1189–1203, https://doi.org/10.1175/JPO-D-16-0239.1.

Wenegrat, J.O., and L.N. Thomas, 2020. Centrifugal and symmetric instability during Ekman adjustment of the bottom boundary layer. *J. Phys. Oceanogr.*, 50, 1793–1812, https://doi.org/10.1175/JPO-D-20-0027.1.

Whitehead, J.A., 1981. Laboratory models of circulation in shallow seas. *Philos. Trans. Royal Soc. A*, A302, 583–595, https://doi.org/10.1098/rsta.1981.0184.

Whitney, M.M., and R.W. Garvine, 2005. Wind influence on a coastal buoyant outflow. *J. Geophys. Res.*, 110, C03014, https://doi.org/10.1029/2003JC002261.

Wiebe, P., R. Beardsley, D. Mountain, and A. Bucklin, 2002. U.S. GLOBEC northwest Atlantic/Georges Bank Program. *Oceanography*, 15, 13–29.

Wijesekera, H.W., J.S. Allen, and P.A. Newberger, 2003. A modeling study of turbulent mixing over the continental shelf: Comparison of turbulent closure schemes. *J. Geophys. Res.*, 108, 3103, https://doi.org/10.1029/2001JC001234.

Wijesekera, H.W., E. Jarosz, W.J. Teague, D.W. Wang, D.B. Fribance, J.N. Moum, and S.J. Warner, 2014. Measurements of form and frictional drags over a rough topographic bank. *J. Phys. Oceanogr.*, 44, 2409–2432, https://doi.org/10.1175/JPO-D-13-0230.1.

Wilkin, J.L., and D.C. Chapman, 1990. Scattering of coastal-trapped waves by irregularities in coastline and topography. *J. Phys. Oceanogr.*, 20, 396–421.

Winant, C.D., 1983. Longshore coherence of currents on the southern California shelf during the summer. *J. Phys. Oceanogr.*, 13, 54–64, https://doi.org/10.1175/1520–0485(1983)013,0054: LCOCOT.2.0.CO;2.

Wise, A., C.W. Hughes, J.A. Polton, and J.M. Huthnance, 2020. Leaky slope waves and sea level: Unusual consequences of the beta effect along western boundaries with bottom topography and dissipation. *J. Phys. Oceanogr.*, 50, 217–237.

Wolanski, E., 1986. An evaporation-driven salinity maximum zone in Australian tropical estuaries. *Estuar. Coast. Shelf Sci.*, 22, 415–424.

Wright, D.G., and K.R. Thompson, 1983. Time-averaged forms of the nonlinear stress law. *J. Phys. Oceanogr.*, 13, 341–345.

Wu, X., F. Feddersen, and S.N. Giddings, 2021. Characteristics and dynamics of density fronts over the inner to midshelf under weak wind conditions. *J. Phys. Oceanogr.*, 51, 789–808, https://doi.org/10.1175/i-D-20-0162.1.

Wunsch, C., 2015. *Modern Observational Physical Oceanography: Understanding the Global Ocean*. Princeton University Press, Princeton, N.J., 512 pp.

Yankovsky, A.E., 2009. Large-scale edge waves generated by hurricane landfall. *J. Geophys. Res.*, 114, C03014, https://doi.org/10.1029/2008JC005113.

Yankovsky, A.E., and D.C. Chapman, 1996. Scattering of shelf waves by a spatially varying mean current. *J. Geophys. Res.*, 101, 3479–3487.

Yankovsky, A.E., and D.C. Chapman, 1997. A simple theory for the fate of buoyant coastal discharges. *J. Phys. Oceanogr.*, 27, 1386–1401.

Yankovsky, E., and S. Legg, 2019. Symmetric and baroclinic instability in dense shelf outflows. *J. Phys. Oceanogr.*, 49, 39–61.

Yoo, J.G., S.Y. Kim, and H.S. Kim, 2018. Spectral descriptions of submesoscale surface circulation in a coastal region. *J. Geophys. Res.*, 123, 4224–4249, https://doi.org/10.1029/2017JC013732.

Young, W.R., P.B. Rhines, and C.J.R. Garrett, 1982. Shear-flow dispersion, internal waves, and horizontal mixing in the ocean. *J. Phys. Oceanogr.*, 12, 515–527.

Zaba, K.D., P.J.S. Franks, and M.D. Ohman, 2021. The California Undercurrent as a source of upwelled waters in a coastal filament. *J. Geophys. Res.*, 126, e2020JC016602, https://doi.org/10.1029/2020JC016602.

Zhang, W.G., and G.G. Gawarkiewicz, 2015. Dynamics of the direct intrusion of Gulf Stream ring water onto the Mid-Atlantic Bight shelf. *Geophys. Res. Lett.*, 42, 7687–7695, https://doi.org/10.1002/ 2015GL065530.

Zhang, W.G., and G.G. Gawarkiewicz, 2015. Length scale of the finite-amplitude meanders of shelfbreak fronts. *J. Phys. Oceanogr.*, 45, 2598–2620.

Zhang, W.G., and S.J. Lentz, 2017. Wind-driven circulation in a shelf valley. Part I: Mechanism of the asymmetrical response to along-shelf winds in opposite directions. *J. Phys. Oceanogr.*, 47, 2927–2947.

# Index

A page number in italics refers to a figure or table.